Hot Air Rises and Heat Sinks
Everything You Know About Cooling Electronics Is Wrong

Hot Air Rises and Heat Sinks

Everything You Know About Cooling Electronics Is Wrong

by Tony Kordyban

ASME Press

Copyright © 1998 by The American Society of Mechanical Engineers
345 East 47th Street, New York, NY 10017

Library of Congress Cataloging-in-Publication Data

Kordyban, Tony, 1957–
   Hot air rises and heat sinks : everything you know about cooling electronics is wrong / by Tony Kordyban
      p.   cm.
   Includes index.
   ISBN 0-7918-0074-1
   1. Electronic apparatus and appliances—Temperature control.
   2. Electronic apparatus and appliances—Design and construction.
   I. Title.
   TK7870.25.K67   1998
   621.381—dc21   98-8455
                                                                                                                                                               CIP

All rights reserved. Printed in the United States of America. Except as permitted under the United States Copyright Act of 1976, no part of this publication may be reproduced or distributed in any form or by any means, or stored in a database or retrieval system, without the prior written permission of the publisher.

ASME *shall not be responsible for statements or opinions advanced in papers or . . . printed in its publications* (B7.1.3). Statement from the Bylaws.

Authorization to photocopy material for internal or personal use under circumstances not falling within the fair use provisions of the Copyright Act is granted by ASME to libraries and other users registered with Copyright Clearance Center (CCC) Transactional Reporting Service provided the base fee of $4.00 per page is paid directly to the CCC, 222 Rosewood Drive, Danvers, MA 01923.

TABLE OF CONTENTS

Chapter 1: We Don't Sell Air 1
Our hero (the author) discovers that his new employer has written some engineering folklore into product design requirements. Should you measure the actual product temperature, or only the temperature of the air coming out of the vents in the back? ***Lesson***: Junction temperature as the source of thermal trouble.

Chapter 2: Every Temperature Tells a Story 9
How hot would a resistor have to be to actually glow? Higher or lower than the melting point of solder? Lab legends always mention glowing components or melting solder, but how hot would that actually be? And what is the ideal serving temperature for ice cream? ***Lesson***: Putting some landmarks on the temperature scale.

Chapter 3: Climate Control Isn't Natural 15
Herbie learns that an environmental chamber is only good for testing products that will eventually be used only in environmental chambers. ***Lessons***: Natural vs. forced convection; thermal runaway.

Chapter 4: Diamond Is a GAL's Best Friend 21
Read the fine print on that thermally conductive epoxy. It may be 50% better than the nonthermal epoxy, but as a conductor it still makes a pretty good insulator. ***Lesson***: Thermal conductivity.

Chapter 5: Lines in the Sand 31
Don't tell a circuit designer which board layout gives the worst thermal performance. He or she will choose it as the only one that will work electrically. ***Lesson***: Introduction to CFD (computational fluid dynamics).

Chapter 6: When Is a Heat Sink Not a Heat Sink? 39
More folklore from EE-land about how aluminum has the magical ability to absorb heat like a sponge and send it off to a parallel universe. ***Lessons***: Convection and surface area; conduction.

Chapter 7: Trade-Offs 47
There are trade-offs between electrical performance, cost, and temperature, so it doesn't pay to be TOO cool. *Lesson*: Junction temperature operating limits.

Chapter 8: Cfmophobia 53
A whole company is infected with fear of Rotary Gas Acceleration Devices (fans). *Lessons*: Fans have a lot of drawbacks that justifiably make people afraid of them, so it is best to plan them in from the very beginning.

Chapter 9: Sieve Cooling System 59
A system that stays cool only because of all of the air leaks that were accidentally designed into the chassis. How to predict the performance of a cooling system that is literally full of holes. *Lesson*: Hand calculation of natural convection flow is nearly impossible to get right.

Chapter 10: Hitting the Wall 67
Natural convection has a limit, because Mother Nature doesn't face much competition and doesn't work hard on process improvement. But computer chips are getting hotter every day. *Lessons*: Natural vs. forced convection cooling.

Chapter 11: Keeping a Cool Head 73
A 25-CFM fan doesn't give 25 CFM of air, and Herbie nearly loses his head over it. Guidance for figuring out fan flow rates is given on the back of a napkin. *Lesson*: The fan performance curve.

Chapter 12: Tempera-Mental Prototypes 79
Cooling an electronic component is different from cooling a power supply is different from cooling a human being. Setting thermal design goals for a project is more than just filling out forms. *Lesson*: Operating temperature limits.

Chapter 13: Misdata 89
Component data books are chock-full of specs that are only valid when they are not important, just like my temperature-telling watch, which is only accurate when it isn't too hot or too cold outside. *Lesson*: Temperature limits in terms of air temperature aren't very useful.

Table of Contents

Chapter 14: Pessimism: A Tool of Quality 95
Herbie and Vlad discover that two fans are not always cooler than one. *Lesson*: Fans in parallel don't always provide redundancy.

Chapter 15: Blowin' in the Wind 101
Junk science and misconceptions in heat transfer. Where does all this baloney come from? Start with the TV weather and the "windchill factor." *Lesson*: Forced convection heat transfer equation.

Chapter 16: Thermocouples: The Simplest Way to Measure Temperature Wrong 109
The most reliable and accurate way of measuring component temperature can also literally explode in your face if you follow Herbie's example. *Lesson*: How a thermocouple works and how it might not work.

Chapter 17: But the Graphics Are Pretty 115
Computer simulation can predict the temperature of electronics even before they are built, and can be wrong to eight decimal places. *Lesson*: More on CFD.

Chapter 18: Too Much of a Good Thing 121
Pin fin heat sinks look like they have a lot more surface area in those magazine pictures. How come they don't work any better? *Lesson*: Convection works with surface area parallel to fluid flow.

Chapter 19: Computer Simulation Software As Test Equipment? 127
Nobody trusts a computer simulation except the guy who did it, and everybody trusts experimental data except the guy who did it. Why not combine the two and get results everybody can mistrust a little? *Lesson*: CFD as a way of interpreting temperature test results.

Chapter 20: Thermotriples 133
Thermocouple folklore and the ongoing debate: Should you weld or solder your thermocouple junctions? It doesn't matter if you're going to measure in the wrong place anyway. *Lesson*: More details of why thermocouples work.

Chapter 21: Mixed-Up Convection 143
Natural convection and forced convection should be friends. Why make them fight each other, unless fans of the Chicago Cubs are involved, and then nothing needs to make sense anyway. *Lesson*: What happens when natural and forced convection work in opposite directions.

Chapter 22: A Dependable Answer 149
How many watts can a 64-lead component safely handle? How big do the vents in my box have to be? What percentage of the heat comes out of the solder side of a printed circuit board? The answer to these and most other common electronics cooling questions is "It depends." *Lessons*: Component package power limits and their limitations.

Chapter 23: Sunscreen or Smokescreen? 155
A university study claims that sunscreen keeps skin 20% cooler than bare skin. This is so obviously wrong that even an EE can spot it. *Lesson*: Temperature is not an absolute scale.

Chapter 24: When 70°C Is Less Than 50°C 161
Is a thermal test done at 70°C and 1000 ft/min air velocity more severe than a test at 50°C and 0 ft/min? Not always. *Lesson*: Convective heat transfer depends on a combination of air velocity and temperature difference, not just air temperature.

Chapter 25: Even a Watched Pot Boils Eventually 165
Roxanne the Intern hasn't learned her cooling folklore. Instead of following the traditional lab procedure of waiting an hour, then recording a temperature, she actually waits until the temperature reaches a maximum, and all heck breaks loose. *Lesson*: Thermal time constants and transient convection.

Chapter 26: The Latest Hot CD 175
When you get a fever, the nurse doesn't have you put some ice under your tongue and then take your temperature again. Herbie wants to put a heat sink only on the components that have been measured as being too hot. *Lesson*: A complex assembly may have more than a single operating temperature limit, and the limit may change under different environmental conditions.

Table of Contents

Chapter 27: What Is a Watt? **181**
How hot does a component get that dissipates 1 watt? Like real estate, it depends on location, location, and location. ***Lesson***: Convection + conduction = conjugate heat transfer, a tricky problem that can baffle your intuition.

Chapter 28: Resistance Mythology **187**
Finding the junction temperature is the key to everything. But it turns out that the only way of calculating it is based on ancient mythology instead of physics. "But all legends have some basis in fact," as Captain Kirk says, so maybe you just stick with the myth until something better comes along. ***Lessons***: Conduction; definition of the thermal resistance between junction and case.

Chapter 29: Thermoelectric Coolers Are Hot **195**
Electrical Engineers love these all-electronic refrigerators. Herbie proposes them for use in a new system until he learns that not only do they cost a lot, but they still require fans and heat sinks, and leave the components hotter than they would be without them. Why are they so bad if they do everything the manufacturers claim? ***Lesson***: Peltier-effect cooling.

Chapter 30: The House of Cards **201**
Even the experts resort to a little mythologizing once in a while. A late night confession reveals that this business of controlling the temperature of electronics to improve performance and reliability isn't nearly as precise as is claimed. There is hope that someday soon the advance of technology might be able to slip a solid foundation under this house of cards without toppling the whole thing. Why isn't anybody worried? ***Lesson***: The not-so-scientific relationship of temperature and reliability in electronics.

Herbie's Homework Helpers **209**
If I have whetted your interest in heat transfer and cooling of electronics, or on the topic that everything you know in general is probably wrong, please go to these sources for lots more detail.

Dedication

To my father, Professor Eugene S. Kordyban, who taught me both reasons why you can't siphon boiling water. And that if you remember the funny reason, you'll never forget the real reason.

Acknowledgments

This book started out as a company internal newsletter on thermal design called *HOTNEWS*, with a total circulation of less than one hundred. Most places wouldn't let you write a report on how the sun comes up in the east without getting sixteen approval signatures. Tellabs not only didn't stop me, but encouraged me to write about design mistakes. They weren't threatened in the least by a humorous style.

Not only that, but management thought the *HOTNEWS* would make a good book for the whole world to read. I am especially thankful to Jim Melsa, a vice president at Tellabs when I started the *HOTNEWS*, now Dean of Engineering at Iowa State University, who pushed me to do this project and who started the ball rolling by helping to find a publisher. Finding the right publisher for such a strange project was accomplished by Ted Okiishi, Dean for Research and Outreach at the College of Engineering at Iowa State. It would have been impossible to write the *HOTNEWS*, and this collection in book form, without the support and creative freedom provided by my manager, Tom Ortlieb, and my director, Paul Smith. They liked the idea of learning from mistakes instead of sweeping them under the rug.

I can't forget to thank Carol Gavin, who worked out the arrangement that allowed me to use material from the Tellabs *HOTNEWS* in this book. Mike Birck, president and CEO of Tellabs, Inc., was very generous in agreeing to this, recognizing the book's potential for educational value and, perhaps, a chuckle or two.

On the home front, I got a lot of moral and material support from Bill Ramsay, Margaret Ramsay, Bertha Thomas, and my sweet wife, Alice Ramsay.

Anonymous thanks go to several *HOTNEWS* readers who suggested, "You're in the wrong business. You should quit engineering and become a writer!" I trust you weren't commenting on my technical abilities.

WE DON'T SELL AIR

CHAPTER 1

> *Our hero (the author) discovers that his new employer has written some engineering folklore into product design requirements. Should you measure the actual product temperature, or only the temperature of the air coming out of the vents in the back?* **Lesson:** *Junction temperature as the source of thermal trouble.*

I first met Herbie in my early days at TeleLeap, Inc., a small company nestled along the Silicon Tollway in the American Midwest. The name TeleLeap comes from the idea that we "leapfrog" current technology to develop new telecommunications products. Some say it comes from our habit of releasing a new product every leap year.

It was that awkward time between learning where the bathrooms were and actually having some work to do. The first job that falls on your desk, and the way you handle it, sets the tone for how the rest of them will go. For better or for worse, the first person to drop into my sparsely furnished office was Herbie.

"So you're the new thermal guy," Herbie said, flopping down into the chair with the broken armrest. "Think you're some kind of *hot-shot* or something?"

I smiled weakly. About three puns can be made about thermal engineering, and in three days I had heard them all.

Hot Air Rises and Heat Sinks

"That's the idea," I said with a shrug. "I'm supposed to be the company thermal guru."

Herbie nodded. "What do you as a guru do?"

"I don't really know yet. I've been brushing up on these heat transfer textbooks."

"Wow, I never saw anybody use a textbook at work. You MUST be a thermal guru."

"Have you got a thermal problem for me?" I asked.

Herbie jumped up to the whiteboard and uncorked a smelly green marker. "Yeah, and here it is in a nutshell. See, we've got this system called Crosser." He drew a box, and then inside of it he drew four vertical lines.

"OK," I said, not much enlightened.

"The Crosser system has been tested and out on the market for about nine months. But now we want to do this —" He drew four more lines sandwiched between the old ones. "We want to double the number of ports in the system."

"Are they going to call it *Double Crosser*?" I asked.

Herbie waved away the suggestion as if it were a mosquito. "Obviously there's going to be twice as much heat in the same space, so the question is, will it still work? Will it get too hot?"

I got up and looked more closely at the problem definition on the whiteboard. From the improved angle it still only looked like a box with eight lines in it. "Maybe you'd better show me the real thing."

Herbie corked his marker and led me to the lab.

TeleLeap looks like any company that makes electronics. It is a chaotic maze of corridors, copy machines, and cubicles crammed with people busy leaving voice mail messages for each other. In some invisible way this all adds up to an organization that makes electronics for the telecommunication industry. TeleLeap makes quite a pile of money on products that 99.9% of the population has no idea even exist.

For example, we sell a lot of Echo Cancelers. Phone calls through satellites used to have an annoying echo, but not after we apply our technology. Who would think you'd need thousands of dollars of electronics to make a phone call sound like a phone call, but you do.

Another problem phone companies have that nobody would ever

imagine is that to save wire they often multiplex several different phone conversations on one wire. Between the time you can say "No" and "Thanks" several other people have already hung up on other telemarketers. There are different, incompatible ways to do this, with cute acronyms like T1 and DS3. When the phone companies need to connect these different lines to each other and still have them make sense, TeleLeap sells them a digital cross-connect system like Crosser.

This means that a lot of the people at TeleLeap are engineers: software writers, electronic hardware developers, and a handful of mechanical engineers. The hardware developers are very talented with their flip-flops and phase-locked loops, but they have only a vague notion that heat is bad for electronics. They want somebody else to get rid of it for them, the way a homeowner calls a plumber to rod out the sewer line. TeleLeap is still not a huge company, and it doesn't need a bunch of thermal rodding specialists, just me. Besides my experience and education in the field, my main qualification for the job was that I was one of the few people in North America willing to do it.

Down the hall was the lab. No test tubes, no bubbling liquids or spinning reels of tape. It was just another office-type room with a few work benches and a row of six-foot tall, steel racks filled with cages of circuit boards. It looked something like Figure 1-1.

Figure 1-1 Heat problems are in the eye of the beholder

Hot Air Rises and Heat Sinks

Herbie swept his hand dramatically toward one of the racks, like a model pointing out a box of Rice-a-Roni on *The Price Is Right*. He said, "This is a 96-port Crosser. What it does is ..." His detailed description has NOT been removed to protect company secrets, but only because I really have no idea what he said. He used a lot of English words such as *megahertz, 16-bit, coaxial cable, time slot interchange,* and *frame relay,* but he might as well have been explaining the rules of Australian football. I smiled and nodded and looked at the pretty blinking LEDs on the front panels.

"... and that's basically all there is to it. What else do you need to know?" he finished.

I probed my hand into a space between a pair of card cages and wiggled my fingers in the warm air and made a concerned face. "Hmm," I said.

"What? Is it too hot?" Herbie said, slightly panicked.

You can't tell accurately how hot something is by feeling it with your hand, but it seemed like something a thermal guru should do. I said, "Is there a temperature test report for this system?"

Herbie tilted his head back and forth as if he were trying to make his brain roll back into its socket. "Oh, sure! The guy who wrote it isn't here anymore, but I'm pretty sure I can find it in the files. He was pretty good. Used to work the grill when we had our department barbecues. Must have been good at thermal if he knew how to barbecue, right?"

The test report was in an old filing cabinet, behind some scope probes. It was only a few pages long, including a drawing of the system and a table of temperature measurements. I took it back to my office for further study.

The report was, to put a good face on it, concise. It forced me to read between the lines. The more I did, the more worried I got. It simply stated that the Crosser System met the temperature requirements contained in the Product Definition Document. I dug up the Product Definition, which referred to a corporate standard, which in turn called out an industry standard. After a day and a half of digging and reading, I thought I had found the problem. Setting a precedent for our future adventures, it was not the original problem that Herbie had first posed.

I was about to find out whether TeleLeap really was like the other companies I had worked for. I had not found a mere mistake in a test report, or a single component that was too hot. What I had seen was a fundamental flaw in the way TeleLeap had been doing thermal engineering, written right into their engineering standards. I didn't have a problem telling anybody — it is my nature to be a big-mouth Mr. Know-It-All — but how would the company react? Would they listen to me? I thought it best to try out my arguments on Herbie first.

"It looks like we've got a lot more work than we thought on this Double Crosser project," I said.

"What do you mean 'we'?" Herbie answered. I later found he was not fond of the phrase "more work."

"Here's the story. You want to know how the Double Crosser would work, based on the assumption that the original Crosser does not have any temperature problems."

"What do you mean 'assumption'? We *know* the Crosser is OK. You have the test report."

"True. But that report has a lot of assumptions in it, too."

"I don't think the barbecue guy could even spell assumptions, much less put any in a report. He was just following the standard."

"That's true, too. And here's what the standard says: *The air temperature rise through the system must be less than 20°C*. That's it. You measure the air temperature at the inlet grill, and the air temperature at the exit grill, and if the difference is less than 20 degrees, you say the electronics are cool enough."

Herbie shrugged his shoulders and said, "So what's wrong with that? We've done that for years. Doesn't it make sense? After all, this is an *air-cooled* system. The air goes in and picks up heat and comes out. The hotter the circuit boards are, the hotter the air is coming out. If the circuits are too hot, the air coming out will be too hot."

"Let's get back to basics. Why do we care how hot the electronics get anyway?"

Hot Air Rises and Heat Sinks

Herbie hesitated, then said, "If a component gets too hot, it'll pop or burn up, or just stop working. Maybe the printed circuit board could catch on fire."

"OK," I said, "and don't forget reliability. Even if it doesn't pop right away, a component could fail in only a short time if it is running hot. By short time I mean before the warranty runs out."

"Oh, no kidding! Those phone company guys are nuts on reliability. They act like if you can't make a phone call, you could die or something."

"Like 911?"

"Um, right."

"So we need to keep the temperature down for reliability. But the temperature of what? The air? Who cares how hot air gets? We don't sell air, we sell electronics. The reliability, and even whether a component will flip its flip-flops, is based on the junction temperature of that component. The junction is the guts of a component where all the action is, you know, the silicon chip. And I don't think measuring the air temperature coming out of our system is a good way of measuring junction temperature."

Herbie still frowned. "I still don't get why. If all of the components were overheating, wouldn't the air coming out be hotter than if the components were all OK?"

"That's only true if all of the components are running at about the same temperature. But my experience is that on a typical circuit board, most of the components are cool, and only a few parts get hot, because they do all the work. When you measure the exit air temperature, you are more or less averaging all of the components together. Compare it to this situation: Suppose you're at the doctor's office waiting room. You have a high fever because you have malaria, but everyone else there is normal because they're just coming in to pick up their nicotine patch prescriptions. The doctor is very busy and doesn't want to look at the whole table of numbers, so the nurse just waves her thermometer around in the air in the waiting room. She has a rule of thumb that says as long as the average isn't over 75°F, then everybody is fine. You'll be sent home without even an aspirin."

"I hate those HMOs!" Herbie said.

"Hold on, now, and pretend you're a microprocessor chip on one of those boards in the Crosser. You're spitting out, say, three watts of heat, and all of the other chips around you are doing only a tenth of a watt each. Do you want me to say your temperature is OK based on the temperature of those puny little chips?"

"No, wait, I don't want to die! I mean — wait, I think I get your point. By measuring only the air, we could be missing some localized hot spots."

"And it only takes one component to fail for the whole circuit board, and maybe the whole system, to fail."

Herbie made a face like he had just dropped his sandwich on the floor, peanut butter side down. "Oh-oh. Looks like WE'VE got a lot more work to do."

I shook my head a little in disbelief. "You buy my story? You're going to change the spec so the temperature limit is based on component temperature instead of air temperature? Just like that?"

"Sure. You're the thermal guru, aren't you?" Herbie said, not even looking at me.

Herbie's reaction turned out to be fairly typical for TeleLeap. People here (and I generously include management in this category) seem to be genuinely interested in hearing about what they are doing wrong, and in figuring out how to fix it (although most prefer to hear the stories about others' boo-boos than their own).

Herbie's explanation for this is, "Relax, bud, it's only *thermal*. It's not like you're talking about something important, like football or the weather."

I have often wondered about the widespread practice of hiding one's mistakes, especially in the world of science and technology. There is no journal called *Hard Lessons Learned*, and I think it is a shame. Because as much as we learn from reading about all of the new discoveries and triumphs of successful projects, we could learn so much more by reading about the things people have done wrong over time. Not to ridicule, or to feel superior, but to save ourselves from the same waste of time and effort. Also, it can be immensely entertaining. Which would you rather see again: a documentary about how the Golden Gate Bridge is built so well, or that film of the "Gallopin' Gertie" Tacoma Narrows Bridge?

Hot Air Rises and Heat Sinks

One main reason there are so many blunders in the field of electronics cooling is that many of the people who get stuck doing it don't have any formal training, and they fall victim to the rich body of technical folklore and superstition floating through the labs of the world. The 20-degree Rule of Thumb for air temperature rise is only one of these widespread fallacies.

The rest of this book is made up of anecdotes, like this one, based on real-life thermal adventures at places like TeleLeap. Learning from mistakes is the central theme, with a sprinkling of heat transfer theory thrown in. Although the heat transfer technology I discuss is real, and the stories are based on real events that happened on the job, for various reasons I have fictionalized many of the details. For example, Herbie himself exists only on a higher plane of reality, being "made up" of my buddies and colleagues, who have each had their turn at his lab bench. Also, the research that TeleLeap is doing into telepathic communication is nowhere near the state of completion that I portray. It seems that sometimes we can only expose the truth by telling artful lies.

CHAPTER 2

Every Temperature Tells a Story

> *How hot would a resistor have to be to actually glow? Higher or lower than the melting point of solder? Lab legends always mention glowing components or melting solder, but how hot would that actually be? And what is the ideal serving temperature for ice cream?* **Lesson**: *Putting some landmarks on the temperature scale.*

I often forgot that not everybody I worked with wore a temperature watch. Sometimes I'd blurt out in a meeting, "Don't worry, that part is only 68°C," and I'd get a roomful of blank stares. It's as if I had just offered to sell them a temperature-measuring watch and told them that it costs "only" 40,000 lire. They had no sense of scale, no idea whether that number was supposed to be big or small, good or bad.

I blamed TV weather, for not converting to the metric system. Herbie blamed me for using "those C degrees that nobody knows what they are." My engineering colleagues just had no familiarity with the terrain of temperature, so I drew for them a map (Figure 2-1), and sent copies of it to everybody in the company mail so they could get familiar with some of its more famous signposts.

Hot Air Rises and Heat Sinks

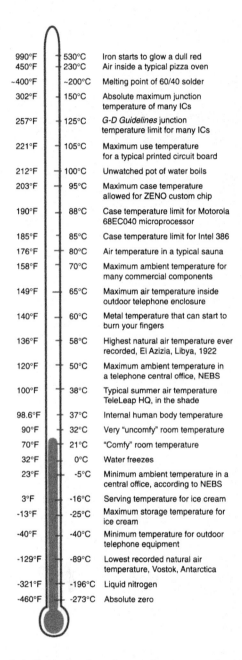

Figure 2-1 Some landmarks on the temperature scale

Significant Reference Points of the Temperature Scale

530°C Herbie talks about seeing the components on his boards "glow red." Think of that as a figure of speech. Plastic melts before it glows, and metals like iron don't even start to glow before they get over 500°C. If you see something glowing on your circuit board, don't call your thermal engineer, call 911.

230°C Pizza oven air. Dough, sauce, cheese, sausage, and 20 minutes produce one of the basic necessities of life.

200°C Solder melts in this range. There are a zillion kinds, all with their own melting points, so don't hold me to it. This is near the top end of our scale, so plenty of other bad things will happen to your components before they de-solder themselves.

150°C/125°C Altera says that the maximum junction temperature for their Programmable Logic Devices is 150°C. At 151°C they might go "poof." The TeleLeap *Good Design Guidelines Manual* says, "OK, maybe, but we'll only trust you up to 125°C."

105°C All the printed circuit boards (PCB) used in TeleLeap products must be made by an Underwriters' Laboratories (UL)-listed manufacturer. UL rates PCBs with a Maximum Operating Temperature. If we use a PCB above its temperature rating, it will not pass UL. If the PCB gets too hot, it becomes soft and warps, and won't maintain the proper distances between traces that are required for safety. Ordinary boards made of glass fiber and epoxy, commonly called FR4, have ratings between 105°C and 130°C, depending on the construction. Boards made of more exotic and expensive materials can go higher.

100°C Water boils. So few people these days know how to boil water that the boiling point doesn't mean much anymore.

95°C Case temperature limit on the ZENO custom chip. Its spec does not guarantee proper operation of this chip if the case temperature (the temperature on the top surface of the component package) is exceeded. This is actually more useful than the specs that limit junction temperature, as it is nearly impossible to measure the

Hot Air Rises and Heat Sinks

junction in operation. This is only one example of a component temperature limit. Others can be higher or lower, but are in this ballpark.

88°C/85°C For comparison, I've listed the case temperature limits for the Motorola 68EC040 and Intel 386 microprocessors. Every part vendor has a different way of stating their limits. Many cop out and give temperature limits that don't mean anything. Also, every component is different — the case limit on one kind of Pentium is only 70°C.

80°C Air temperature in a sauna. We had to come a long way down the scale before we found a temperature that human beings can withstand. Humidity in a sauna is very low, about 5%, so you can maintain your body temperature (see 37°C) by sweating. Otherwise, 80°C would kill you in a matter of minutes.

70°C This is the mythical maximum ambient temperature for most commercial components. Refer to Chapter 13, Misdata, for an explanation of why this rating is more than useless.

65°C The telecommunications industry standard for outdoor equipment enclosures says the air temperature inside the box should not exceed 65°C, no matter what: sunshine, heat wave, nuclear testing.

60°C A metal surface at this temperature will probably raise a blister on your skin if you have to touch it for any length of time. Poor heat conductors like wood or plastic are a little safer because they don't transfer their heat energy into your flesh as fast. They need to be 10 to 20 degrees hotter to start toasting you.

58°C Hottest weather on record, from El Azizia, Libya.

50°C Learn this one temperature if you intend to stay in the telecommunications business. This is the maximum ambient temperature for telephone central offices, according to the industry standard, Bellcore's *Network Equipment and Building Standard* (NEBS). See Herbie's Homework Helpers for a complete description.

38°C The proverbial "100 degrees in the shade," which is the mandatory Midwest air temperature if you have to get all dressed up to attend an outdoor wedding.

Every Temperature Tells a Story

37°C Body temperature, on the inside. This is analogous to your junction temperature, if you consider the place where your mom used to measure your temperature when you were a baby to be your junction.

32°C A very sweaty room temperature.

21°C A comfortable room temperature. Naturally, my office was kept at 16°C, and I had to wear a sweater in the summer, even when it was 32°C outside.

0°C Freezing point of water. I reminded Herbie that a 50/50 blend with antifreeze will freeze when it is twice as cold as this. A week later he actually tried to figure out what I meant.

–5°C The minimum ambient temperature for telephone central offices set by the industry standard. It seems to me that a building full of running equipment will hardly ever get this cold, even if Ma Bell runs low on firewood for the stove.

–16°C Typical temperature of ice cream when it is served. Ice cream does not have a single melting point, being a mixture of many substances.

–25°C If you store ice cream below this temperature, ice crystals don't grow, and it will keep indefinitely. Sorry, but your home freezer probably doesn't go this low.

–40°C This is the minimum air temperature expected for an outdoor equipment enclosure. It is also the only place on the scale where Farenheit and Celcius agree.

–89°C Coldest weather on record, from Vostok, Antarctica.

–196°C Liquid nitrogen boiling point. There is a recipe on the Web for making instant ice cream using liquid nitrogen. Yum.

–273°C Absolute zero. If we could achieve this temperature in the lab, hell would freeze over, and the Cubs could finally win the World Series.

Hot Air Rises and Heat Sinks

Learn these landmarks, these signposts, and these reference points well. It will help if you take this map with you to a pizza joint, study it while you chow down on a deep-dish with everything, and polish it off with some ice cream. The second-degree burn on the roof of your mouth from the pizza and the ice cream headache will help to impress this temperature scale permanently onto your brain. You will never again embarrass yourself by telling your friends you saw a glowing electronic component (except for LEDs and tube filaments).

If your thinking still has not been Celsified, then memorize this conversion equation:

$$°F = (9/5) \ °C + 32 \qquad (2\text{-}1)$$

CHAPTER 3

CLIMATE CONTROL ISN'T NATURAL

> *Herbie learns that an environmental chamber is only good for testing products that will eventually be used only in environmental chambers.* **Lessons**: *Natural vs. forced convection; thermal runaway.*

I was making thermocouples in the lab that I share with Doc, the Environmental Engineer. Trimming, twisting and soldering a couple dozen pairs of fine-gage wires was tedious, so I had an old tape playing on the boom box. As fate would have it, the tape was the Firesign Theatre comedy troupe playing a guy buying a used car with "climate control." Only this option does a little bit more than defrosting or air-conditioning. When he selects the "Land of the Pharaohs" setting, he is instantly transported to North Africa in the middle of some alternate-universe version of World War II. Just then, Herbie walked into the lab, looking to schedule some time in the Enviromatic 9000.

"Sorry," I said, "Doc isn't around."

Herbie decided to wait for her. "I've got a question for you anyway," he said, "It's about this test we do in the Enviromatic 9000. How much margin do you think we really ought to have?"

"Margin?" I asked.

Hot Air Rises and Heat Sinks

"In my department we have this traditional test. We've done it as long as I've been here, except that I can't find a document anywhere that says exactly how or why you're supposed to do it. And because it isn't written down, if your circuit board doesn't pass it, you start to argue about how much margin you're really supposed to have. How much margin is enough in your opinion?"

"Tell me about the test," I said.

"All of the circuit boards for the Crosser II are supposed to work up to 50°C, so we run them in an environmental test chamber at 70°C for 24 hours. If they work OK, then we know we have a 20 degree margin. But sometimes a board poops out at 68°C or 65°C. Then the arguing starts. Since it really only has to work up to 50°C, and it works all the way up to 65°C, isn't that good enough? Isn't that enough margin?"

I made a face as if I were thinking very hard, holding up one finger to keep Herbie silent. The Firesign Theatre finished off their parody of "God Bless America," then I shut off the boom box, smiling. I said, "You do it in this environmental test chamber right here, the Enviromatic 9000?"

"Yeah," he said, "Or sometimes in our small one back in our lab."

The Enviromatic 9000 is a room-size steel box with triple-insulated walls and a huge door like a meat locker (Figure 3-1). With its refrigeration and heating coils, and a steam injection system, it can produce any air temperature and humidity condition from –65°C to 150°C, and 0 to 100% relative humidity. It is a fairly standard hunk of test equipment you can find at any large electronics facility, especially one that makes things for the military.

"And these boards are all cooled by natural convection?" I asked.

"They're convection cooled."

I made the face Mrs. Kratch used to give us in the second grade when we said *unbrella* instead of *umbrella*. "Convection is a general word," I explained, "Like *quality*. When SchlockMart claims they sell a quality product, it means nothing, because they don't say whether it is good quality or bad quality. Convection means giving off heat to a moving fluid, like air. If the air is being pushed along, such as by a fan, then we call it forced convection. Natural convection is

Climate Control Isn't Natural

Figure 3-1 Enviromatic 9000 test chamber set for "Land of the Pharaohs"

when the motion of the air is caused by the heating itself. A hot object, like this soldering iron, heats up the nearby air. The hot air rises and the surrounding colder air flows in to take its place, and the whole thing keeps going by itself. Put your hand right here and you can feel the hot air rising."

"OK," Herbie said. "There aren't any fans in the Crosser II, so I guess you'd call it natural convection."

"Then I guess the size of the margin isn't really important," I said, "because your test is bogus. It is worse than doing no test at all."

"Bogus? How come you never just answer the question I ask you?" Herbie said, looking at the floor, then at the ceiling. "OK, why is our test so bad? I thought you'd be impressed that we were testing our boards with any kind of temperature margin at all."

I finished soldering the last thermocouple bead. "Stick around and we can do a little demo, using this power supply board I'm supposed to start testing anyway." With Herbie's help, I glued the temperature sensor to a diode on the power supply board, plugged the board into its test fixture, and wheeled the whole thing into the Enviromatic 9000. The test fixture was set to draw the full rated load of the power supply so that its components should be working under the

Hot Air Rises and Heat Sinks

worst possible conditions. The chamber was programmed to keep the air at a steady 50°C. After about an hour, the chamber control panel displayed 50°C air temperature, and my thermocouple meter showed a rock-steady diode temperature of 71°C.

"These results look pretty good," I said. "Sam, the power supply engineer, told me that the operating limit for this diode is 95°C. Above that and it starts to go into thermal runaway. So far it looks like we have a 24°C margin of safety. But watch what happens when I do this —"

I pushed the OFF button on the Enviromatic 9000. The lab suddenly became quiet, as its giant blowers and compressors shut down. The air temperature inside held at 50°C because of the chamber's heavy insulation, but almost immediately the diode temperature began to creep up. We both stared at the red LED display as the temperature went to 75°C, then 80°C, and higher.

"What's going on?" Herbie asked. "How come the power supply keeps getting hotter? It was maxed out until you turned off the chamber controls."

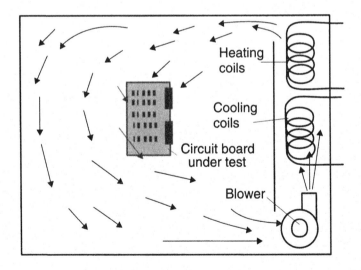

Figure 3-2 "Environmental" chambers aren't very good at creating a natural convection environment

I said, "Chambers like this, including the little one in your lab, are designed for a specific purpose — to get to the set-point temperature as fast as possible, and then hold that temperature very tightly. The typical way to do this is for the chamber to have a powerful blower that drives the chamber air through a nest of heating and cooling coils. When it gets too hot, the cooling coils kick in. When the air temperature drops, the heating elements pop on. The blower keeps going constantly, recirculating the air inside the chamber and keeping it well mixed. You can't run the coils without the blower because the coils would get damaged. So the only way to shut off the air flow is to shut off everything, and then you can't guarantee temperature regulation. The blower doesn't make a gentle breeze, either. It's enough to blow off that clip-on tie you keep in your desk for important presentations." (See Figure 3-2 for a tie-blowing velocity distribution.)

Herbie said, "So while the chamber is ON, the blower is blowing on the power supply. That's like having a fan cooling it off. And when you shut off the chamber, you shut off the blower, and it gets hotter, like it would when the customer uses it, since there isn't any fan in the system. Hokey smokes! How big a difference can that make?"

The diode temperature hit 89°C. I said, "You're getting an idea of it right now. It depends on a lot of factors, such as how fast the air is blowing on your board, but forced convection can be from 10 to 100 times more effective at removing heat from components than natural convection. So it doesn't seem quite fair to say you design a product to work in natural convection at 50°C, and then do all your high temperature tests with a blower. See why your test is bogus?"

"When will the diode stop getting hotter?" Herbie said.

"I don't know. Either it will reach equilibrium soon, or it goes into thermal runaway. Some types of components, like diodes, change their characteristics with temperature. This one increases its power dissipation as the temperature goes up. After a certain point, the hotter it gets, the hotter it gets, until something fries."

A few minutes later, the diode hit 97°C, then rapidly went up to 108°C, and suddenly started to go down. The test fixture showed output at 0 volts and 0 amps. "Thermal runaway," I said, snapping off

Hot Air Rises and Heat Sinks

the input power to the test fixture. "I'm sure Sam will be thrilled to find out he was right."

Herbie plopped down on one of the lab stools. "What would have happened if we tested this power supply according to the traditional margin test? Would we have found the thermal runaway?"

"Maybe not," I said. "If you did it in the Enviromatic 9000, like we did here, probably not. With the blower on, and the chamber set for 50°C, the diode was about 71°C. If you increased the set point of the chamber to 70°C, you would probably add the same 20°C increase to the diode, so it would be only 91°C. That isn't high enough to trigger the thermal runaway. So you probably would have approved this power supply for production, assuming it had a 20°C margin of safety. And the first time somebody runs it at full load at 50°C, it could destroy itself."

Herbie laid his head on his arms on the lab bench and moaned, "So what are we supposed to do now? Should we test all the boards at 80°C? 90°C? How much margin do we need in the chamber to prove we can work in natural convection at 50°C?"

I sighed. I wished we had that "climate control" from the Firesign Theatre and could set it on "The Land of Natural Convection." But there just isn't any such chamber. So instead I said, "Stop thinking in terms of margin, and try to make the test look a little bit more like the way the customer is going to use the product. Stop testing naturally cooled systems with the blower on. These test chambers just aren't very good at reproducing a natural convection environment. And I suppose we should sit down and write these things up in a formal test spec, or procedure, or something."

"Or at the very least," Herbie said, brightening up a little, "we should supply an Enviromatic 9000 with every Crosser II, so the customer can install the Crosser inside it. At least we know it works well there."

DIAMOND IS A GAL'S BEST FRIEND

Read the fine print on that thermally conductive epoxy. It may be 50% better than the nonthermal epoxy, but as a conductor it still makes a pretty good insulator. **Lesson**: *Thermal conductivity.*

Herbie, smiling smugly, placed a small vial on top of my copy of the long-awaited annual Extended Surfaces Issue of the *Journal of Electronic Packaging*.

"I just saw that rep from Chro-Sink-TempCo, and I convinced him to let me have a sample of this stuff. It's going to save us on the Satan chip problem," he said.

"How did you convince him," I asked, "by letting him buy you lunch?"

Herbie burped indignantly. "Hey, we talked about stuff at that lunch! Work-type stuff. Anyway, I told him about the Satan problem and by the time dessert came, he figured out what we needed, and luckily he had a sample in his case to let me try out."

Satan was the code name for a custom chip for some new secret project being conducted in Arizona. Since the perfection of the miniaturization process (thanks to Irwin Allen), it was possible to cram all of its circuits into a package the size of one of those fake press-on fingernails. All those gates flipping and flopping at 666 MHz

Hot Air Rises and Heat Sinks

generated just a little too much heat for the package to get rid of — about 5 watts. Herbie and I were struggling to find a way of mounting a heat sink on the Satan. There just didn't seem to be enough room for a heat sink big enough to get rid of 5 watts.

"It's called *High Thermal Conductivity Epoxy*," Herbie went on. "Sure, it takes 24 hours to set, has a useful shelf life of only 5 days, and costs an arm and a kidney, but the rep says it has a thermal conductivity 68% higher than that heat sink glue we already use. You know how they do it? Silver! It's epoxy filled with silver powder. That's why it's so expensive."

I went to the whiteboard and erased the table analyzing the defensive match-ups between the Bulls and the Knicks. "Let's see how much saving this silver-plated mucilage can produce. Give me the data sheet."

"This is the equation for how heat conducts through a solid. **Q** is the amount of heat, **t** is the thickness of the glue layer, **A** is the area of the glue joint, and **k** is the thermal conductivity of the glue. Δ**T** is the temperature difference from one side of the layer of glue to the other" (Figure 4-1).

Herbie said, "I can see how this makes sense. The thicker the layer, the harder it is for heat to flow through it. And the bigger the

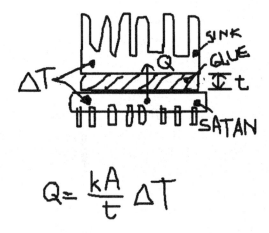

Figure 4-1 A sticky barrier between Satan and keeping cool

area, the easier it is, because the heat doesn't have to squeeze through a small opening. And the higher the conductivity, the lower the temperature rise, which is good. We want the temperature rise to be really small, so the component temperature goes down."

"OK," I said, "let's see what the difference is between my ordinary glue and your High Conductivity glue." We calculated:

	Old Glue	Silver Glue
Q	5 W	5 W
t	0.005 in.	0.005 in.
A	0.28 in.^{2}	0.28 in.^{2}
k	0.0168 W/in./°C	0.0282 W/in./°C
ΔT	5.3°C	3.2°C

"The silver does help," I said, snapping the cap on the whiteboard marker.

"Yeah, but only by about 2 degrees," Herbie said, "and we need to get the temperature down at least 35 more degrees."

Herbie was falling into the common trap of hoping for some Thermal Magic Wand that could solve all his heat problems, the equivalent of the fabled 200 mile per gallon carburetor. This kind of magical thinking says that aluminum soaks up heat like a sponge and makes it disappear, and that Styrofoam is a miracle because somehow it knows to keep hot things hot and cold things cold all by itself. You may have heard of some of these magic wands and wondered why you don't see them in use: thermally conductive epoxy or grease, heat pipes, thermoelectric coolers, air-powered vortex refrigerators, piezoelectric vibrating fans, liquid nitrogen spray guns, special paints that enhance thermal radiation, and flexible bags of fluorocarbon liquids. All of these things actually exist, but they don't do anything for free. You can use an exotic material or a fancy gizmo to move heat from one place to another, but in the end, you still have to take the heat out of the component and dump it into the surrounding air, and there is only so much heat you can move for a given temperature difference.

For the Satan, the problem was still how to get the heat out of the package and into the air. The most direct way is to increase the

Hot Air Rises and Heat Sinks

surface area with a heat sink. If we couldn't fit a simple heat sink in the available space, we couldn't fit a magic wand in there either.

Herbie would have been even more disappointed with his silver-filled epoxy if I had shown him Table 4-1. Instead of calling their product *High Thermal Conductivity Epoxy*, Chro-Sink-TempCo should have called it the *Slightly Less of a Thermal Insulator Epoxy*. It is like Herbie and I arguing which of our salaries is closer to Bill Gates'. Sure, one of us makes more money, but I don't think Bill is looking over his shoulder to see if either of us is gaining on him. Compare the following materials.

Table 4-1

Material	Thermal Conductivity (W/in./°C)
Air (not moving)	0.00076
Nylon	0.00635
Heat sink glue	0.0168
Brick	0.0175
Glass	0.02
Herbie's glue	0.0282
Alumina	0.7
Steel	1.7
Silicon	2.5
Brass	3.05
Aluminum	5.5
Gold	7.4
Copper	10.0
Silver	10.6
Diamond	16.0

Right away you notice that air by itself is not a very good conductor of heat. That is why you want to have really good contact between a heat sink and the body of the component. It would be 23 times better to have a thin layer of glass between them than an air gap of equal thickness.

Another thing that becomes obvious is that the heat sink glue and Herbie's glue both fall in the range of materials that anybody would

Diamond Is a GAL'S Best Friend

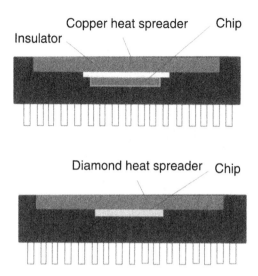

Figure 4-2 A chip can be directly bonded to a diamond heat spreader without shorting out

call good thermal insulators. Not until you get to alumina in the table do you start to think of the materials as being actual conductors instead of impeders of heat transfer. That is why the name *thermal epoxy* seems like an oxymoron to me.

At the bottom of the table is one of those magic wands that even we Thermal Gurus have been lusting after for many years. It is an unfortunate coincidence that most of the best heat conductors are also good electrical conductors, like aluminum and copper. Much of the time that doesn't matter, but there are some very important applications where it would be nice to have a material that is an excellent conductor of heat, and an excellent insulator of electricity.

For example, the chip inside a Gate Array Logic (GAL) device can actually be quite small compared with the size of the outer package (Figure 4-2). The chip is also where the transistors and diodes are, which is also where all the heat is generated. The plastic encapsulant doesn't conduct heat very well, and the chip can get very hot. When this happens, the manufacturer might insert a heat spreader into the package. The heat spreader is a slab of aluminum

Hot Air Rises and Heat Sinks

or copper to which the chip is bonded. To prevent the parts of the circuit from being shorted together, they have to put an electrical insulator between the die and the heat spreader, which partly defeats the purpose of trying to conduct the heat to the outside of the package.

Diamond has just the properties we want. It is one of the best heat conductors around, and it is an excellent electrical insulator. It also is very durable and can withstand extremely high temperatures. If you had a slab of pure diamond as a heat spreader, you could etch the circuits on its surface without any insulator and heat would shoot right out of the device like funny sounds out of a whoopee cushion under compression. We dream of someday having whole circuit boards made of diamond.

There are still a couple of minor barriers to achieving this new thermal magic wand technology. You'd probably get only 20 or 30 heat spreaders out of the Hope Diamond, and then what? Start in on the Crown Jewels of England?

It isn't as bad as that. Physics and chemistry are coming to the rescue again. Those guys in the white coats have already figured out how to make artificial slabs of diamond film by vapor deposition. For high-power applications, they have built prototypes of actual circuits on diamond film substrates. Someday soon you will see DeBeers get into the integrated circuit business. We'll have circuit boards with gold contact connectors and diamond-encrusted components, and I'll still hear from Marketing that we can't add a fan to the system because it is too exotic.

Herbie's Cheat-Sheet on Thermal Conductivity

Herbie complained that heat conduction always seemed confusing to him, especially the units of conductivity. Why were they in watt/meter/°C or BTU/hour/feet/F? Why was it so complicated? In college you heard that heat conduction was just like electricity conducting through a wire. Except in electricity, you usually talk about *resistance*, not conductivity. And resistance was a simple, ordinary unit — ohms, not coulombs per fortnight per league, for example.

So I told Herbie a story about the Joule monkeys living on islands in the Fourier Sea. These monkeys can't swim, and they hate being crowded together. They always try to make the biggest space between themselves and the other monkeys. Whenever they see an opportunity to move to a less crowded place, they run for it. And the bigger the difference in how crowded it is, the faster they run to get there. You could even write a little mathematical relation that says the speed that the Joule monkeys run is proportional (that's what $\propto$ means):

$$\frac{Joules}{time} \propto (Crowdedness_1 - Crowdedness_2) \qquad (4\text{-}1)$$

Once in a while, a palm tree gets top-heavy with monkeys trying to escape the crowded beach and falls over, bridging the water between two islands. This time it happens to fall on a neighboring island that has very few monkeys per acre (if you define crowdedness as the number of monkeys divided by the available living area, and these monkeys do). Joule monkeys scamper across the fallen tree to enjoy the new relative lack of their companions.

How fast do the monkeys cross the tree? Knowing the monkeys' aversion to crowding, we can write an equation for the monkey-crossing rate. First of all, the monkeys continue to avoid crowding, even while crossing the tree-bridge (unlike people, who will jam together for short periods to get through doorways and such). So the wider the tree, the more monkeys can cross at the same time without becoming more crowded than they are on their home island. They won't get on the bridge if it is more crowded than the island where they already are (Figure 4-3).

$$\frac{Joules}{time} \propto width \qquad (4\text{-}2)$$

We already know that the monkeys run faster if the difference in crowdedness is bigger from one spot to the next. They have more urge to run fast if the distance between the crowded place and the sparse

Hot Air Rises and Heat Sinks

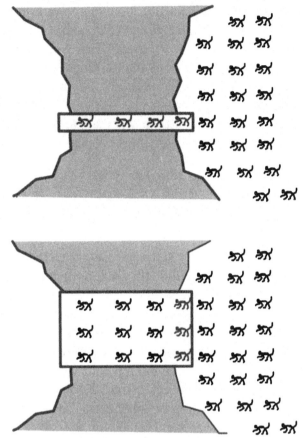

Figure 4-3 The wider the bridge, the more Joule monkeys per second can cross

place is short. So the longer the tree, the slower the monkeys run across it. That is because the shorter the distance between the two places, the easier it is for each monkey to sense the difference in crowdedness.

$$\frac{Joules}{time} \propto \frac{1}{length} \quad (4\text{-}3)$$

The third thing that matters about the tree is how easily the monkeys can get footing. It might have thorns or be very slippery, which would slow down their crossing. The monkeys have given numerical ratings to all their trees called *klamberability* (*k* for short). The higher the number, the easier it is to run across.

$$\frac{Joules}{time} \propto klamberability \qquad (4\text{-}4)$$

Putting all these together gives you the complete equation for monkey transfer:

$$\frac{Joules}{time} = \frac{klamberability \times width}{length} \times (Crowdedness_1 - Crowdedness_2) \qquad (4\text{-}5)$$

This is exactly how conduction heat transfer works, if we connect up the right parts of the monkey business to the physics. The Joule monkeys are units of heat energy — let's call them Joules (a convenient coincidence). The more Joules that are crowded together into an object, like an electronic component, the higher its temperature. Temperature is how we sense the amount of heat energy stored per molecule of an object. Roughly speaking, a large component with one Joule stored in it has a lower temperature than a small component with one Joule stored in it. Heat energy, or Joules, have the same aversion to crowding as the monkeys (and I) do, so it tends to flow away from high temperature (crowds) to low temperature (secluded areas).

Let's say an aluminum bar falls down, connecting our island component and another island of nice, cool air (Figure 4-4). What determines how fast the heat can travel across the bar?

The speed of heat flow through the aluminum bridge depends on the same things as the monkey-transfer speed: the difference in temperature (crowdedness), the width of the path (in this case, the cross-sectional area of the bridge because heat flows through the material, not just along the top), and the length of the bridge. It also depends on what material was used to make the bridge. Some materials are good conductors, like aluminum and diamond; some are worse, like paper and plastic and glue. The better the material conducts heat, the higher its value of *k* (for conductivity — that's right,

Hot Air Rises and Heat Sinks

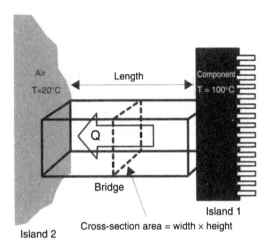

Figure 4-4 Heat crossing a bridge to find some breathing room

scientists can't spell any better than monkeys). So the equation for heat crossing a bridge looks like this:

$$\frac{Joules}{time} = \frac{conductivity \times area}{length} \times (Temperature_1 - Temperature_2) \quad (4\text{-}6)$$

If you remember, energy per time is power, and a Joule per second is actually a watt. We like to talk in watts instead of Joules, because then we can compare the electric power in the components with the heat flow rate in watts. For heat flow rate we use the letter Q, named after nobody in particular. Now you and Herbie can understand conduction, if you remember the crowd-hating monkeys:

$$Q = \frac{k \times A}{l} \times (T_1 - T_2) \quad (4\text{-}7)$$

If you play with this equation, you can figure out why k has units of watt/meter/°C.

$$k = \frac{(Q \times l)}{A(T_1 - T_2)} \text{ or } k = \frac{(watts \times meters)}{meters^2 \, (C_1 - C_2)} \text{ or watts/meter/°C} \quad (4\text{-}8)$$

LINES IN THE SAND

CHAPTER 5

> *Don't tell a circuit designer which board layout gives the worst thermal performance. He or she will choose it as the only one that will work electrically.* **Lesson**: *Introduction to CFD (computational fluid dynamics).*

In June, Herbie called. "We're at the high-level design phase of the PTM," he said, "and we're looking for input from you on the thermal design requirements." The PTM was the Phone Tag Module, a new feature being developed for our old workhorse PBX system, at the request of a major customer. You know how annoying it is playing "phone tag," getting a voice mail message, calling back and leaving a message, then receiving a return message, over and over. The PTM automates the process — the two voice mails communicate directly with each other without bothering you. All your phone calls are promptly returned, whether you ever get them or not, which in some businesses is the whole point.

"Right now we don't know much about this module. We just want to put a Rule of Thumb in the Design Description Document, something for the designers to shoot for," he said.

I hate Rules of Thumb. They are like drawing lines in the sand. As soon as you draw one, somebody wants to step over it. But it was June, and a lunch hour pick-up basketball game was forming, so I

Hot Air Rises and Heat Sinks

thought up a Rule of Thumb anyway. "This PTM goes into the old PBX system?" I asked.

"Yes."

"How much power does the hottest board in it use now?"

"Five watts."

"Put that in your DDD. The PTM shall not use more than 5 watts power. Then it should be pretty safe from a thermal standpoint, since it won't be any worse than the old boards, which are OK."

Herbie was pretty happy with that answer — until September.

In September, Herbie caught me in line at the cafeteria. "What if the PTM is more than 5 watts?" he asked. "I've added up the components we've identified so far, and maybe it'll be a little more than 5 watts — 6, 7, maybe 10 watts at the outside."

I said, "If you like, bring the Bill of Materials and a drawing of the PTM layout to my office, and we can do a detailed thermal simulation and predict component temperatures."

Herbie stole a grape out of my fruit salad, waving off my suggestion. "No, no, no. Nothing that complicated. Just another Rule of Thumb like the last one, something simple we can keep in mind, but, just a little higher. Five watts wasn't really an absolute limit, right? It was based on what other boards in the system were getting away with. Maybe there is a lot of margin, and we could easily go up to, say, 10 watts?"

"OK," I said, not able to afford any more of my lunch being nabbed. "There's a Rule of Thumb for power density in natural convection. Take the total power of the board and divide by the surface area of the front and back of the board. If that number is below 0.1 watts per square inch, then you should be OK."

Herbie wrote the number on his hand. "0.1," he said. "Got it. We'll stick to this limit, no problem."

In October, Herbie came to my office bearing a copy of the Bill of Materials for the PTM. "What about 0.12?" he asked, "that's just a little over 0.1 watts per square inch. Is it still going to work at 0.12? The custom chips came in a little higher than anybody expected." He showed me the list of parts for the board, along with their power dissipation.

Lines in the Sand

"Look, Herb," I said, "why do you bother asking me for these thermal design rules? You don't follow them anyway. These Rules of Thumb are very crude, and so they have to be very conservative. If you are trying to do a really optimal design, they aren't of much use. It's like trying to find a street address in downtown Chicago using what they taught you in the Cub Scouts about how moss grows on the north sides of trees and flowing water leads to civilization."

"So what am I supposed to do?" Herbie said. "I don't want to do what we used to always do — build the prototype and see if it is too hot. I want to make sure it will work before I build one."

"I'm glad to hear you say that. That is 90% of the problem. Here's a treat for being a good little design engineer," I said, handing him a gumball from the jar on my desk. "Now here's what we ought to do. About 6 months ago my boss coughed up the money so we could buy a CFD analysis program called The ThermaNator. I have been doing a lot of playing around with it, and I think we can use it to predict component temperatures."

"CFD. I like it. Sounds great. What's CFD?"

"It stands for computational fluid dynamics," I said.

I explained how The ThermaNator works. Back in the 1960s, the first computer codes were written to solve the extremely painful equations of fluid flow and heat transfer. It took PhDs to define the problems, and supercomputers to crunch them to a solution. It was so labor intensive and expensive that only the most high-budget, high-priority problems could take advantage of it — like the space program, or the optimization of golf ball dimples. But now that a car radio has more computing power than Apollo 11 when it landed on the moon, CFD has become practical on the desktop computer.

The software has come a long way since those days too, now having a nice graphical front-end for defining the problem. All you have to do is draw a picture on the computer of your circuit board and indicate with a few mouse clicks the locations of heat sources, the material properties of the board and components, and the shape and locations of the air vents. These are called boundary conditions. If there is a fan, you specify how fast it blows as a boundary condition. After you set up the problem, ThermaNator divides up the three-dimensional space

Hot Air Rises and Heat Sinks

you drew into thousands of little pieces, and for each piece, or grid cell, it writes a set of energy balance and mass balance equations (remember the conservation of mass and energy?). If you set the boundary conditions right, the program then has a jillion equations and a jillion unknowns, which it proceeds to solve by the highly advanced method of trial and error. It guesses a solution, plugs it into the jillion equations, and compares the answer to the original guess. Then it adjusts the guess and does it again. And again. And again. Until the next morning, when it guesses close enough and announces that it has converged on a solution.

The solution is the air velocity and temperature of every grid cell, which is displayed as a bunch of little arrows representing air flow, and color contour lines showing temperature. We can use software like this to predict the temperature of components on a circuit board that is cooled by airflow, if we know their location and power. It can even calculate natural convection, which is a particularly tricky thing to do by hand.

I showed Herbie one of the fancy color temperature maps on my computer screen. It was a test board I had built in my lab out of 10 resistors, just to see how well ThermaNator would predict its temperature. "I made the computer model first, then built and tested the board in my lab. The resistor temperature predictions come out within plus or minus 5°C," I said proudly. "That's pretty good compared with almost any other way of doing it. Give me a layout of your PTM and I'll run it through The ThermaNator."

Herbie chewed his gumball thoughtfully. "Are you sure you can't just give me a number? How about 0.15?" he said. "The problem is I don't have any kind of layout to give you right now. I have a 90% done schematic, and this list of the major components, but I haven't even started to think about what goes next to what."

I said, "Great! How about this? Give me the list of components and I'll make up my own layout and see how it turns out. If I make the worst possible thermal layout and it still has OK temperatures, then you know the board will be OK no matter how you design it. If I find some layouts that don't work, then you can use that as a design rule for how NOT to lay it out."

Herbie shrugged, which I took as a green light to begin the project.

The PTM is cooled by natural convection, which means the air flows from bottom to top driven by the heat of the components. The air gets hotter from bottom to top. So design rule No. 1 is: To keep things cool, place them near the bottom edge of the board. Design rule No. 2 is: Spread heat sources as far apart as possible. Turning these rules upside down, I came up with the worst possible layout I could think of. I crammed the highest-power parts together near the top edge of the board.

I let The ThermaNator crunch on that overnight, and it gave me the component temperatures in Figure 5-1.

Unfortunately, some of the components in my worst possible design were too hot. The voice memory modules were rated by the manufacturer to operate only up to 85°C, and the custom chip I stuck between them was supposed to stay below 100°C. So I went back to following the design rules and came up with a better layout, just to make sure the PTM had at least one design that could work.

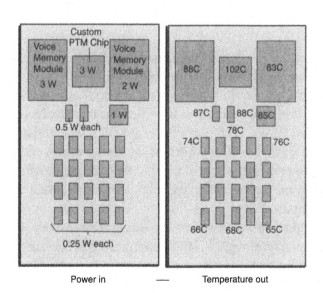

Figure 5-1 The worst possible layout for the PTM bunches all of the high-power components together near the top edge of the board, and some parts end up too hot

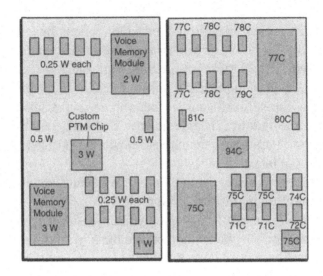

Figure 5-2 This layout spreads the power apart and gives lower component temperatures — it may not be the optimum layout, but we could live with it

The ThermaNator analyzed the layout in Figure 5-2 and gave me a somewhat improved temperature map.

I felt pretty good giving Herbie the ThermaNator results. I gave him copies of the two pictures and even wrote in fat, red marker on the first one, "BAD." That gave him some very specific design advice. "This circuit board is on the border between passing and failing thermally. There are some ways of placing the parts that will make them too hot," I said, "so you will have to be careful to spread out the high-power parts."

In November, Herbie brought me the finished layout. He handed it to me sheepishly, saying, "Could you maybe run this through The ThermaNator again, before we send it out to be built?"

I nearly shouted, "I don't have to! This is almost exactly the layout I told you NOT to use."

"Yeah, what a coincidence," Herbie chuckled nervously. "Actually, we were having a heck of a time coming up with a layout that would work electrically and have good routing. Then we looked at your picture, the one marked "BAD," and suddenly everything just fell into place. The board practically routed itself after that. You have a real gift."

"Oy!" I said. "It's another example of Murphy's Law applied to thermal/electrical design. The best layout for electrical performance is usually the worst for thermal performance. It's not just because nature is perverse — there are real reasons for it to be true. The higher frequency stuff is usually the hottest, but it also requires very short lead lengths between components, so naturally all of the hottest parts want to be very close together. The thing I'm kicking myself about is that if I had let you come up with your own layout, you probably wouldn't have found the absolute worst thermal layout, and your layout might even have had OK temperatures. But because I told you what won't work, you just absolutely have to have it."

"It *is* perfect for us," he admitted.

We haggled for a few days, moving parts around and running more thermal simulations until we found a compromise that worked both thermally and electrically. It was a good first exercise for using thermal simulation to aid board layout. We learned lots of lessons.

One I won't soon forget is to let the electrical engineers do the layouts from now on.

WHEN IS A HEAT SINK NOT A HEAT SINK?

CHAPTER 6

> *More folklore from EE-land about how aluminum has the magical ability to absorb heat like a sponge and send it off to a parallel universe.* **Lessons**: *Convection and surface area; conduction.*

You know a heat sink when you see one. A heat sink is that chunk of aluminum that looks like a really thick comb (Figure 6-1). The heat sink and the fan are the two most basic tools of the electronics cooling engineer, and they are the most widely misunderstood. We'll talk about the fan mythology later. There is enough mystery surrounding the properties of aluminum to keep us spellbound.

"I don't know why it works," Herbie told me once. "Maybe you can explain it. Ever since I was a kid soldering together bootleg cable TV decoders, I was taught, if a part is getting too hot, slap an aluminum heat sink on it and it'll cool off. What is it about aluminum that cools things off?"

"There is no such thing as a heat sink," I said.

"No such thing? Then what are those things you're always telling me to clamp onto my hot diodes?" Herbie asked.

What you call it depends on your definition.

One way of defining something is to make a list of all the things that it is not. Here are some examples of heat sinks that are not being heat sinks.

Hot Air Rises and Heat Sinks

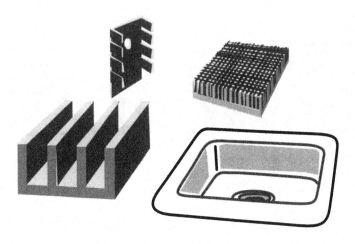

Figure 6-1 Some commonly found sinks

Real Life Example #1: The Cheapo Disk Drive Cover

A while back I did a temperature test on a disk drive, the kind you see in a personal computer (Figure 6-2). We were considering it for use in one of our products. When the disk was accessed, a couple of components on the circuit board mounted on the top of the disk drive became quite hot.

The vendor was very clever and made a couple of dimples in the sheet metal cover over this board so that the cover would come into contact with these two components, acting as a heat sink for them. This is an example of when a cover is not a cover, which doesn't necessarily make it a heat sink, either. A quick push of a finger on the cover revealed that somebody had missed a dimension someplace — there was a noticeable gap between the dimples and the tops of the components. The dimples not only wouldn't help, but probably made things worse by choking off airflow over the components. This cover/heat sink was no heat sink at all.

The term *heat sink* must have been coined by an electrical engineer. It is too close in concept to the *current sink* used in circuit theory classes. The term *sink* is not to be found in heat transfer textbooks. When they mention those aluminum combs, they call them *extended*

When Is a Heat Sink Not a Heat Sink?

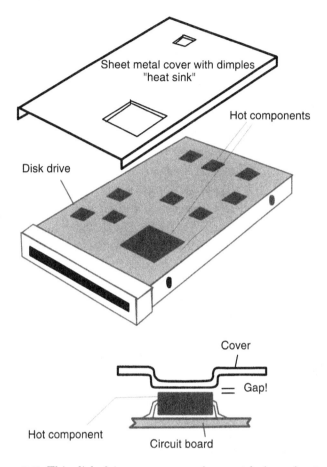

Figure 6-2 This disk drive cover wound up not being a heat sink

surfaces. In this sense, a heat sink is NEVER a heat sink. Why?

An electrical current sink is a dot on a schematic into which current flows endlessly and never returns. Such a thing doesn't exist in the real world, but it is a convenient concept to use when drawing a circuit. A sink is a place we can dump something and then pretend it no longer exists, such as the kitchen sink. We dump our greasy water down the drain and forget about it. But like that dirty water that shows up again on somebody's beach, heat that flows into a heat sink doesn't just disappear, and we ultimately can't forget about it.

Hot Air Rises and Heat Sinks

An ideal heat sink is a thing that absorbs heat endlessly without getting hot. This is impossible. Heat is energy, and the Conservation of Energy is not just a good idea, it is the Law. Heat can't flow forever into a chunk of aluminum and just disappear, or migrate into a parallel universe. If you put heat into a chunk of anything, either the heat will flow out into something else or the chunk will get hotter. Technically, its temperature will increase.

You add a heat sink to a component because the component is too hot. How does that aluminum comb make it cooler? In the usual case, the ultimate heat sink (the place we want all the heat to go) is the roomful of air. Heat energy moves only from high temperature to low temperature, unless you give it a shove. Electrical energy enters your component, flips some bits around, then changes to heat energy. That heat energy increases the temperature of the component. Then it moves through the surfaces of the component into the air. The small surface area of a tiny component limits its contact with the surrounding air, which means there will be a large temperature difference between the component and the air. The heat energy just can't squeeze through that small surface area fast enough, and your component gets too hot. You can think of it as smoke building up in your kitchen while you're burning some bacon on the stove. The bigger the open window, the faster the smoke can get out. If the smoke can't get out as fast as it is being generated in the frying pan, the smokier the kitchen gets (the more heat backing up in the component, the higher its temperature).

So you attach your *extended surface* to the top of your component. The reason it looks like a thick comb is that all those fins pack a lot of extra (or extended) surface area into a small volume. Heat energy now flows out of your component into the extended surface device, which has much more surface in contact with the air than the original component package. You are opening the window wider. The heat can get out into the air much more easily, and so the temperature of the extended surface, and of the attached component, goes down.

The heat energy has still not disappeared. It is in the air. But keeping the room cool is somebody else's problem and you can go to sleep at night thinking the heat really did go into a parallel universe.

A simple equation describes this whole process (Figure 6-3):

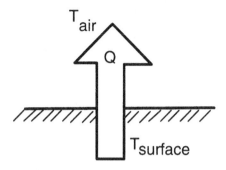

Figure 6-3 Heat flows out the window of surface area

$$Q = h A (T_{surface} - T_{air}) \qquad (6\text{-}1)$$

Q = the rate of heat energy flowing out of your component, expressed in units of energy per time, or watts

h = the convection coefficient, which depends on how air is flowing nearby

A = the surface area in contact with the air

T$_{surface}$ = temperature of the solid surface (your component)

T$_{air}$ = temperature of the air

The whole idea of a "heat sink" is to increase the value of **A** so that the difference between **T**$_{surface}$ and **T**$_{air}$ can go down.

Real Life Example #2: The First Attempt

On an earlier version of the same disk drive from Example 1, there was no sheet metal cover. Glued onto one of the hot components was an aluminum plate, about ⅛ inch thick, with exactly the same outline as the top surface of the component (Figure 6-4). It had no fins or pins or *extensions* of any kind. Was this heat sink a heat sink? (Hint: The disk vendor next tried replacing it with a bigger piece of metal.)

Hot Air Rises and Heat Sinks

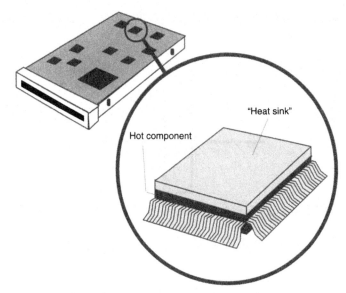

Figure 6-4 Guess how much improvement in temperature this heat sink gives

Real Life Example #3

There is a company that actually makes heat sinks out of plastic. They don't work very well, but they make a great teaching tool. By my previous argument, they should work fine because they can have lots of surface area, sometimes even more than the standard heat sinks made out of aluminum and copper. Plastic is much easier to mold into intricate three-dimensional shapes. What I didn't explain about Eq. (6-1) is that $T_{surface}$ is the temperature of the surface in contact with the air. We have been assuming that the solid is all one temperature, which isn't always so. Aluminum is a good conductor of heat, plastic is poor (about 1/1000 as good as aluminum). Heat conduction (see Herbie's Cheat-Sheet at the end of Chapter 4) through a solid has an equation with a similar form:

$$Q = k\ A/L\ (T_{hot} - T_{cold}) \qquad (6\text{-}2)$$

k = the conductivity of the solid
L = the length of the heat path through the solid from hot to cold

When Is a Heat Sink Not a Heat Sink?

Let's practice with this equation on the heat sink from Example 2. Its one redeeming feature is that it has very simple geometry, which makes conduction easy to calculate (Figure 6-5). Let's say our component is giving off 1 watt of heat, and it has to pass through the ⅛-inch thick heat sink before it comes in contact with the air. We want to know whether an aluminum or plastic heat sink will work better.

Heat Sink Material	Aluminum	Plastic
Q	1 W	1 W
k	180 W/m/°C	0.2 W/m/°C
A	0.00065 m²	0.00065 m²
L	0.0032 m	0.0032 m
$T_{hot} - T_{cold}$	0.027°C	24.6°C

The aluminum heat sink has a temperature rise from the surface in contact with the air to the surface in contact with the component of much less than 1°C. If we made the same heat sink out of plastic, the surface touching the component would be almost 25°C hotter than the surface in contact with the air. This temperature rise adds to the temperature of the component.

For example, let's guess a value of **h** to be 20 W/m2/°C so we can figure out our component temperature. The air temperature is a

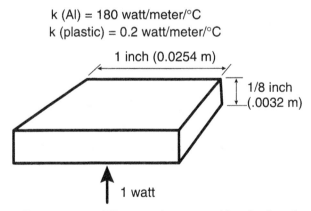

Figure 6-5 Temperature difference from one side of a flat plate to the other depends on the thermal conductivity

Hot Air Rises and Heat Sinks

comfortable 25°C. For both the plastic and the aluminum heat sink, the temperature rise from the air to the top surface of the sink is

$$T_{surface} - T_{air} = Q/(h\ A) = 1W/(20W/m^2/°C \times 0.00065\ m^2) = \mathbf{77°C} \quad (6\text{-}3)$$

Add that to the air temperature and you get a surface temperature for the heat sink of 102°C. Add that to the temperature rise through aluminum from the table and you still have 102°C for a component temperature. But the component with the plastic heat sink is 24.6°C hotter, or about 127°C.

The situation gets worse as the heat path (**L**) gets longer. You can imagine the difficulty with trying to conduct heat through long, narrow fins.

Other Real-Life Examples I Have Seen

1. Pin fin heat sinks jam-packed with pins. They have plenty of surface area, but the pins are so close together that air has a hard time flowing through them. Good airflow is to an extended surface what music is to ballet (imagine ballet without music).

2. Similar to tightly packed pins is the common mistake of not aligning the fins of a heat sink with the direction of air flow. In natural convection, the air always flows upward (barring centrifugal forces and artificial gravity). Yet often I see heat sinks with the fins running horizontally — turned 90 degrees to the flow. If the flow doesn't go through the fins, it doesn't contact the surfaces, and the whole thing is a waste of good aluminum.

3. A heat sink is not a heat sink when it is a branding iron. Make sure very hot extended surfaces won't be touched by people who don't deserve to look like a hot dog fresh off the grill — 70°C and up is too hot to touch for extended periods of time.

There is no end of defining what a heat sink is not. I am afraid the term *extended surface* is not going to catch on, so even I will have to keep on using the name heat sink. I hope everyone will understand why I wince every time someone says it from now on.

TRADE-OFFS

CHAPTER 7

> There are trade-offs between electrical performance, cost, and temperature, so it doesn't pay to be TOO cool.
> **Lesson**: Junction temperature operating limits.

After I had been at TeleLeap about a year, my boss called me into his office. It was time to be "PIMPed," as we called the Process Improvement Management Process. Because, technically, I reported to the Quality Simulation Department, my boss wanted to set some "metrics" for how I was doing my job as a Thermal Guru. Metrics was the latest Quality fad. Somehow you had to quantify everything you did on the job, and then work to improve those numbers. Naturally, I was as enthusiastic as a groundhog crossing Highway 34 out in front of the building on payday at quitting time.

"How about this?" Al suggested. "If you're doing your job of getting everybody to improve their thermal designs, every year the average component temperature of TeleLeap products should go down. That's an objective metric, even if it might be hard to determine."

Maybe I should have been a lawyer, because I saw a loophole in his idea right away. "But that's not how I see my job. I don't want the products to be cooler just to be cooler. I think the average component temperature might go up a lot every year if I'm doing my job right."

"How can that be?" The bait was taken, the hook was set, and I began to explain.

Hot Air Rises and Heat Sinks

Reducing component temperature costs something: not only directly (heat sinks and fans and making holes in chassis cost money), but in other facets of the product design. One board might be split into two to spread out heat. Two devices might share a load to prevent overheating when one could perform the same function by itself. Clock speed might be reduced to keep power dissipation low. Some product concepts may not appear feasible because they are too hot to fit in the allotted space. On a new Crosser board, an expensive switching power supply may replace a linear regulator that is throwing off too much heat and giving a hard time to neighboring devices. There are many trade-offs to be made when playing the component temperature game.

In an ideal design world, I don't try to reduce component temperatures as much as possible. I want them to be as hot as ALLOWABLE, because that means that other facets of the design have been given as much freedom to be optimized as possible. Simply put, don't force components to be cooler than they have to be.

On a typical circuit board that we make, about 95% of the components dissipate little or no heat, and they are way below their operating temperature limits. The other 5% are the ones that keep me in business. So if I am doing my job right, and I advise the designers of the boards so that cooling resources (heat sinks and the like) are used only where they are needed, and every component heats up to just below its operating temperature limit, then 5% of components will go down in temperature and 95% will probably go up. If TeleLeap optimizes its total product designs (not just component temperature), the average component temperature across all product lines should go up year after year.

Al didn't like the idea of tying my PIMP rating to increases in component temperatures, so the idea of metrics was tabled. Since then, I have had two more bosses, and they haven't tried to put me on the metric system yet.

How Hot Is Too Hot?

I am not saying turn up the juice and fry those babies. Not at all. But don't reduce temperature just to be cooler. Find out the maximum operating limit of each component, and stay just under it.

Trade-Offs

Herbie, who once took a free, all-day seminar on electronic reliability (including lunch) sponsored by a heat sink vendor, would say, "But every 10°C increase in temperature cuts component life in half! Keep everything as cool as possible to maximize reliability!"

That Rule of Thumb was probably never true. It comes from the white-coated world of chemistry, where there is a general principle that chemical reactions go faster the higher the temperature. Years ago the military adapted that concept to predicting how temperature makes electronic components fail. They gathered tons of questionable data from the field, then correlated the data with this iffy assumption about chemical reaction rates and came up with the military handbook on electronic reliability (MIL-HDBK-217). It quickly became an industry standard because they wrote the use of it into all the procurement contracts for military hardware, so everybody knows it by heart. MIL-HDBK-217 is the source of the myth that component failure rates double with every 10°C increase in temperature. But most people don't remember that even MIL-HDBK-217 states that long-term nominal operating junction temperatures lower than 70°C have ZERO effect on reliability. So spending money or other resources to reduce junction temperature below 70°C will buy you nothing. The truth is that the temperature that starts hurting a component may be even higher than that. But it is different for different kinds of components.

What are these maximum operating limits? Here is one of the big secrets of the electronics industry that is rarely published. Everybody agrees that for every component there is some temperature above which it should never be operated, if you want it to last longer than the warranty period. But every person has a different idea of what that temperature is. The general approach taken by most buyers of electronic components is to ask the vendor to give a temperature limit. You might recall that a typical digital semiconductor device has an "Absolute Maximum Limit" from Texas Instruments or National of 150°C. Then we "derate" this data sheet value by some safety factor, or factor of ignorance, and come up with something a little lower, such as a maximum junction temperature for digital integrated circuits of 110°C. Table 7-1 gives reliability temperature limits from the

49

Hot Air Rises and Heat Sinks

TeleLeap Good Design Guidelines Manual. It was cobbled together from a hodgepodge of more or less respectable sources, but it is not necessarily the last word on temperature limits.

Table 7-1
Operating Limits for Good Reliability

Type of Component	Maximum Temperature
Bipolar transistor	125°C junction
Field effect transistor	125°C junction
Thyristor, SCR, TRIAC	125°C junction
Diodes, except LED	125°C junction
LED	110°C junction
Linear semiconductor	105°C junction
Digital semiconductor	110°C junction
Hybrid semiconductor	110°C junction
Complex, LSI, VHSIC, VLSI, microprocessor	125°C junction
Memory	125°C junction
Capacitor	Max. ambient rating −10°C
Composition resistor	Max. ambient rating −30°C
Film resistor	Max. ambient rating −40°C
Wirewound accurate resistor	Max. ambient rating −10°C
Wirewound power resistor	Max. case rating −125°C
Thermistor	Max. ambient rating −20°C
Potentiometer	Max. ambient rating −35°C
Inductor	Hot spot rating −15°C
ILD (Injection Laser)	110°C junction
APD (Avalanche Photo Diode)	125°C junction
SAW	125°C junction

Also, keep in mind what maximum operating limit means. It DOES NOT mean that you can run an LED up to 110°C junction when the board is sitting on your bench in the (sometimes) air-conditioned lab. It means that you should design everything so that under worst-case operating conditions (maximum ambient and altitude, board in the worst possible slot in the system, and the system loaded with all of its other hottest boards, plus whatever other realistic bad things might happen, such as actual signals passing through the circuit), the LED junction doesn't go over 110°C.

By the way, for some LEDs, the worst case is not a normal operating condition — if the LED is an alarm light that only comes on in an emergency. In this case, you'd have to simulate the emergency to get a real idea of the maximum temperature. In general, LEDs are not thermal problems (but relays that turn them on have been).

The numbers in Table 7-1 are intended only to shoot for the longest possible life. They don't say anything about the *function* of the component. High temperature may change the signal timing or other functional parameter enough to make the circuit start hiccuping long before reliability becomes a question. The functional operating limit of a component may be even more difficult to pin down, because it may depend totally on the intended circuit. A crystal oscillator that drifts 10% from its nominal frequency with temperature isn't necessarily a functional failure, unless your circuit needs a very accurate clock.

Lately I have been toying with a new kind of metric for my PIMP. It would be based on that little home invader, Goldilocks (of Three Bears fame), who might say, "Those circuits were too hot, and those circuits are too cold, but these TeleLeap circuits are JUST RIGHT!"

CFMOPHOBIA

CHAPTER 8

> *A whole company is infected with fear of Rotary Gas Acceleration Devices (fans).* **Lessons**: *Fans have a lot of drawbacks that justifiably make people afraid of them, so it is best to plan them in from the very beginning.*

Herbie and I went out to lunch, someplace fancy, with metal forks and such. He was on a job rotation in the Marketing Department, so he did lunch a lot. "And another thing," he said, "that comic strip Dilbert isn't as funny as it used to be."

The first thing he had been complaining about was me pushing him to use fans in the new project he was assigned to. The ComComCon project included some new hardware, but it was mostly a marketing concept for a form of cut-rate phone service. The phone customer, in exchange for a much lower rate, would have to listen to commercials during his or her calls (the first "Com" in Commercial Communication Connection). Local calls would have a 30-second ad at the beginning of each 10-minute segment, and long-distance calls would have a 60-second ad every 5 minutes, at least to start. Phase I of the project was developing the hardware to interconnect the ad-playing machine to the telephone switch. Phases II and III had plans for targeting specific ads to individual callers, and the feature of allowing receivers of such calls to block ads, for a small fee. Sponsored

Hot Air Rises and Heat Sinks

Figure 8-1 Cfmophobia

communication was a huge, untapped market, according to Herbie's new colleagues.

"In the world of telecom, fans are bad," Herbie said. "You said so yourself, in writing no less, right there in the Thermal Design section of our new official corporate document, *The Good Design Guidelines*."

"You just have Cfmophobia," I replied.

"Gesundheit."

"A lot of folks at TeleLeap have it. Fear of CFM, or cubic feet per minute, the stuff that comes out of fans. Fans aren't bad. They do one thing really well, which is move air around. It's just that they are tricky."

"Tricky is bad."

"Which is trickier," I asked, "designing custom VLSI chips or putting a fan in a sheet metal box? Which one has more risk, and which one do we do every day?"

No answer came from Herbie.

"They way I like to put it," I said, "is that TeleLeap is pushing the envelope of natural convection. The world of commercial electronics cooling has gone on ahead to thermostatically controlled multispeed fans, heat pipes, thermoelectric coolers, high pressure jet-impingement, boiling fluorocarbon baths, etched silicon passage liquid-to-liquid heat exchangers, helium-filled spring-loaded microscopic copper pistons, and diamond film heat spreaders. And we get the shivers when someone suggests putting a fan in a box."

Herbie protested, waving a fork full of lettuce, "But fans fail! They seize up. They make noise. They fill up the electronics with dust. And then you need a filter to keep the dust out. And if you have one fan, you need two, so that when the first one poops out, you have the spare to keep you going until you can fix the first one. So then you need a sensor to tell when the fan fails, and an alarm to ring when the sensor detects something, and software to detect and report the alarm, and a procedure for responding to the alarm, and then a service guy —"

" — or person —"

"— or person has to go to the site and replace the fan and clean the filter."

"I'm glad you actually read that section in the *G-D Guidelines*."

"That's in the *G-D Guidelines*? It's just what we in Marketing always say at the design review when somebody says maybe we should put in a fan. They taught me that the very first day, right after they showed me how to wear suspenders properly. The real reason we don't want fans is because Joe Blow at PhoneComTechCo won't buy our stuff if it has a fan and the competitor's gizmo doesn't."

I said, "A guy with the name Joe Blow should probably like fans. But you don't have to fling your ranch dressing at me. I agree with everything you're saying. Don't use fans if you don't need them, because they do have some drawbacks. But I need to convince somebody at TeleLeap, and it might as well be you, Herbie, that fans are not automatically bad. And even if they are bad, they will soon become a necessary bad thing. Fans are bad the way the lifeboats on the Titanic were bad: they cluttered up the decks and blocked the view out of some of the portholes. My bet is that fans will become necessary sooner than anybody thinks, which will come at a bad time such as at the very end of a design project. A month before the Absolutely Last Final Review (ALFR), we'll find out something is too

Hot Air Rises and Heat Sinks

hot, and after a long meeting, we'll finally cave in and decide to add fans to the system. And because fans are tricky, it'll be too late in the project to do them right. So either the schedule will be blown, or the fans will be stuck on, like duct-taping a plastic garbage bag over a broken window on a Rolls Royce, and they won't work right, we'll have to do it all over, and then everyone will say, 'See, fans are bad.'"

Herbie smiled and shrugged innocently, "What makes you think things are going to get so hot all of a sudden? So far we've been able to get along by poking a few extra air holes here and there, and maybe adding a heat sink or two to really hot components."

"Which way are all the trends in electronics going? Surface mount — more. Trace widths and spacing — skinnier. Lead spacing — finer. Clock speeds — way up. Gates on a chip — seems to be doubling every month. What happens when the clock speeds go up? Lead lengths have to go down, which means all the components need to be closer together. Higher clock speeds mean more electrical noise, which means more shielding and smaller air holes. Fiber optics are popping up everywhere now, putting these amazingly hot lasers right on the board. And customers are expecting telecom equipment to start shrinking the way mainframes squeezed down to the size of a VCR. All of this adds up to more watts in less space, and eventually you cross the border where natural convection doesn't go anymore. I'm seeing parts from Motorola that say right on their data sheets that they are designed to work only with fan cooling — they don't even consider natural convection anymore. It may not be long before we won't be able to buy the latest, high-tech devices that can be cooled by natural convection. Remember what happened when we did that cost-reduction project on the RoadRunner modem chip? The vendor repackaged the die from a Pin Grid Array, about 2 inches square, into a dinky fine-pitch surface mount flat pack about one-eighth the size. That saved a lot of board space, except — because we didn't have fan cooling — the heat sink wound up being bigger than the old Pin Grid Array!"

Herbie was beginning to lose his appetite. "I don't know why you're so upset. The ThermaNator simulation you did on the ComComCon said that it didn't need any fans."

"In Phase I. But what about Phase II and Phase III? I heard that next year they want to cram another 250 watts into the chassis."

"Yeah, but I just need to get past the Phase I ALFR. By next year I'll be rotated back to my old department in Engineering." He laughed, and then the lesson began to strike home. "But by then, the chassis design and the board spacings and the power supply will all be frozen. If they decide to add a fan in Phase II, it'll wind up being a real kludge. And Phase II boards won't work in a Phase I chassis, because it won't have fans. Hmmm."

"You were right," I conceded. "Fans are tricky. They probably only have a good chance of working right if they are designed into the system from the beginning."

Herbie said, "And if you have Cfmophobia, it'll be impossible to do that, since at the beginning, you don't even want to think about fans."

I was so pleased by the concerned look on his face that I picked up the check.

CHAPTER 9

SIEVE COOLING SYSTEM

A system that stays cool only because of all of the air leaks that were accidentally designed into the chassis. How to predict the performance of a cooling system that is literally full of holes. **Lesson**: *Hand calculation of natural convection flow is nearly impossible to get right.*

Every year the company that developed The ThermaNator — Temperatures Limited, Ltd., holds an international users' conference. It is their way of rewarding their customers, who spend 10-hour days in darkened offices staring at flickering computer screens of thermal simulation software, by inviting them to some exotic location, like London, Las Vegas, or Paris, to spend 12-hour days staring at flickering slide projection screens in a darkened hotel meeting room.

In case you forgot, ThermaNator is the thermal simulation software that I use for predicting airflow and temperatures. One year, the users' conference was held in London. I convinced my boss to let me go by writing a paper on how I used ThermaNator to analyze why component temperatures weren't changing as fast as they were supposed to in an environmental chamber. He bought it, so I had a week of warm beer and stops at the British Museum. Plus, of course, the users' conference.

Hot Air Rises and Heat Sinks

During a break at the conference, a ThermaNator staffer backed me into a corner and asked me whether ThermaNator was actually saving TeleLeap time and money. He was being a good salesman. He wanted a real-life endorsement to tell a prospective customer. I started my canned speech about my environmental chamber paper, but he wasn't impressed. Then I told him a good (and true) story about the value to TeleLeap of a good thermal simulation program.

About a year before, InteleLeap, our international division located in Iceland, sent me a prototype shelf full of circuit boards for thermal testing. The code name for the product was "Boris," a ringing voltage power compensator designed for use in phone systems in former Soviet countries. Apparently, when the KGB went kaput, they didn't bother to remove the tremendous number of wiretap devices installed in the network over the decades. The wiretap devices draw their power directly from the phone line. On some lines there were so many bugs that it was hard for the ring generator to produce enough power to ring the actual phone on the end of the line. Boris is designed to detect and compensate for this "secret" load, even though all it really does is buzz unmanned sets of earphones all over Moscow.

Boris looks like a typical telecom equipment shelf (Figure 9-1). The boards plug vertically into a backplane, and air flows by natural convection through slots in the bottom and top. The shelf mounts in a standard telephone rack and can be stacked up to seven high. I measured the component temperatures in the single Boris prototype. All of the component temperatures were OK, for a shelf running *by itself*.

"Thank you very kindly for your test report," Helga from Iceland said, "but could you now tell us what happens when you stack Boris seven shelves high?"

"Do you have six more shelves to send me?" I asked.

"Sorry, no. Isn't there some way of calculating what happens when you stack the shelves?"

I stalled by telling her the joke about the infrared camera and the traveling salesman. The problem Helga wanted me to solve sounds easy. You know what one shelf does, because you've already meas-

Sieve Cooling System

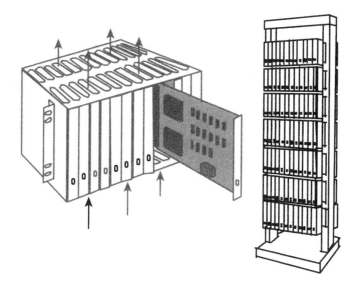

Figure 9-1 How hot does a Boris shelf get when it is stacked up in a rack

ured it. It should be simple to figure out what two shelves do when you stack them — just add the two shelves together — right?

Natural convection is more complicated than that. Each shelf on the stack adds two effects: first, the extra boards add heat, which generates more airflow; second, the friction and blockage of the extra boards tends to slow down the airflow through the stack. It is not easy to predict from looking at the boards which of these opposing effects is stronger. Therefore it is hard to know before you add a shelf to the stack whether you will be increasing or decreasing the airflow.

Don't forget that the air also gets hotter as it passes up through the stack of shelves. The air that exits the first shelf becomes the inlet air for the second shelf. So every time a shelf is added, the airflow could go up or down, and the air temperature increase through the stack depends on the airflow rate. It is the interplay between the extra heat, the extra blockage, and the temperature rise through the stack depending on the airflow that makes the whole business

Hot Air Rises and Heat Sinks

tougher to predict than the price of TeleLeap stock just before the quarterly earnings are announced.

I proposed a simple way to estimate the effect of stacking shelves: I measured the air temperature going into the shelf and the air temperature coming out. On average, the temperature rise was about 13°C. Then, I held the shelf up to the light, looked through the air holes, and pretended that I was an air molecule. It didn't *seem* like there was a lot of flow blockage, so I thought it was fairly conservative to say that the airflow added by the heat of each shelf would probably balance the flow blockage added by each shelf. An answer that at least had the benefit of being easy to compute: air temperature should go up 13°C in every shelf.

So if the room air were 20°C, air exiting the bottom shelf would be 33°C, and leaving the second shelf 46°C, leaving the third shelf 59°C, and so on. Because component temperature depends on the local air temperature in the shelf, it would increase from shelf to shelf in the same way. A relay in the bottom shelf that is 90°C becomes 103°C (90°C +13°C) in the second shelf, and 116°C in the third — OOPS! Too bad. That relay isn't supposed to go above 105°C, so that means you can't stack more than two Boris shelves high, by this simple method of calculation. After two shelves you would need to insert a baffle, which would start the stack over with fresh, 20°C inlet air (Figure 9-2).

Helga did not like this approach, no matter how much work it saved me, because it meant that InteleLeap could fit only five Boris shelves in a rack instead of seven, because of the space wasted on baffles. She also pointed out that Boris didn't put out any more heat than a typical ring generator shelf, and we routinely stacked them seven high with no baffles. Mysteriously, these were OK thermally, because they had been tested extensively in the past and they even used the same relay.

So I abandoned the simple approach and gave ThermaNator a try. I used it to simulate a single Boris shelf and tweaked the simulation parameters until the temperatures predicted by the simulation matched the temperatures I had measured in the real Boris shelf. Then, using those same parameters, I simulated a stack of seven

Sieve Cooling System

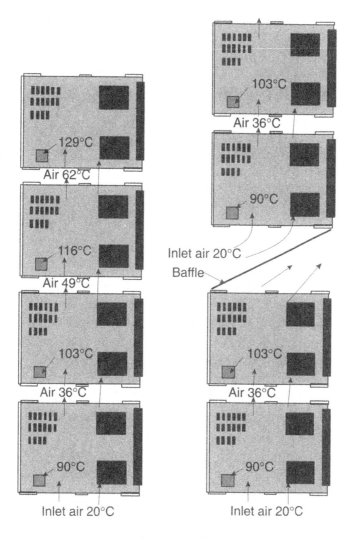

Figure 9-2 According to the simple method, all temperatures go up 13°C per shelf, which forces an air baffle to be inserted every two shelves to keep the relay below 105°C

Boris shelves. Table 9-1 gives ThermaNator's predictions of how component temperatures increase from shelf to shelf.

Hot Air Rises and Heat Sinks

Table 9-1
ThermaNator Results for Boris Stack-Up

Shelf	Exit Air	Relay	Airflow Relative to Shelf 1
7	42.2	97.2	2.43
6	41.8	97.2	2.31
5	41.1	96.4	2.22
4	40.1	95.9	2.12
3	38.8	94.4	1.96
2	36.6	91.9	1.68
1	34.1	89.2	1.00
	20.0	⇐ Inlet air	

Look at Shelf 1. They are numbered from bottom to top, just like in the table. The exit air is about 14°C hotter than the inlet air, which agrees pretty well with the air temperature rise of 13°C I measured on the real shelf. But what is happening in the rest of the shelves? The air only goes up another 2.5°C from Shelf 1 to Shelf 2! That's a lot different from what I expected based on my first approach. My simple method said everything should go up by 13°C per shelf. By the time you get up to Shelf 7, the change in exit air is less than half a degree!

Start comparing the relay temperatures from shelf to shelf and you see the same thing. They also do not go up 13°C per shelf. In fact, in Shelf 7, the component temperatures don't go up any more at all. How can this be?

This is the Sieve Cooling System at work. I have thought about patenting it, but the lawyer said you can't patent leaks. What I haven't told you yet is that my simulation included one additional small detail: a 1-inch air gap between the top of each shelf and the bottom of the next. This is actually how our racks are built, and it has long been suspected that these gaps might be an important factor in cooling. But I could never prove it, because the slow speed of natural convection flow is nearly impossible to measure and it is not

Sieve Cooling System

even practical to tell whether air is flowing in or flowing out of these gaps. It could be simultaneously going in some places and coming out others for all I knew.

But ThermaNator was able to calculate these flows through the gaps, even though my simple approach ignored them. Look at the column for relative airflow. I left out the units because I only want to compare the flow from one shelf to the next. What this tells you is that the flow rate through Shelf 7 is nearly 2.5 times the flow rate through Shelf 1. And where did all of this extra air come from? It had to leak into the stack of shelves through the gaps. That's why I call it the sieve cooling system. It depends on leaks to work.

My initial hand calculations assumed that the airflow through each shelf was identical. I had to assume this, because there was no way to guess how much air might leak in or leak out of each gap. I had to ignore what turned out to be a very important part of the cooling system, and that's why my simple way of predicting component temperatures was so far off from reality.

The ThermaNator simulation did not have to ignore the gaps between shelves, and therefore demonstrated its true value. I already knew that a stack of seven ring generator shelves would work thermally, because I had measured component temperatures in real shelves like that many times. From previous tests, I knew air temperature did not build up from shelf to shelf equally, but I had no way of explaining it until ThermaNator predicted it. From its maps of airflow, at last I had a clear explanation of *why* it works the way that it does. It's the difference between knowing that flight is possible and being able to design your own airplane. You've seen birds fly — you know they can do it. But only after you understand how flight works can you make things that aren't birds fly too. It allowed me to take the previously unquantifiable sieve cooling system and apply it to another design.

"So did this save your company millions of dollars?" the salesman asked.

"That's not the point," I said. "I never bothered to add up the savings, and I don't think it even helped to speed up the project. What we gained was an understanding of how the complicated process of

Hot Air Rises and Heat Sinks

natural convection in a leaky stack of shelves works. That by itself is worth a lot."

"So you wouldn't say a million dollars, then, was saved?"

"Oh, less than that, I think," I said.

The salesman grinned, satisfied, and wrote "nearly one million dollars saved" in his leather-bound notebook. He thanked me and headed off for a well-earned cup of tea.

Hitting the Wall

CHAPTER 10

Natural convection has a limit, because Mother Nature doesn't face much competition and doesn't work hard on process improvement. But computer chips are getting hotter every day. **Lessons**: *Natural vs. forced convection cooling.*

Near the end of the ThermaNator Users' Conference in London, a few of us Thermal Gurus toddled off to a nearby pub to quaff a pint or two of ale. Those of us from the telecom and military hardware industries were ragging on poor Clement, who worked in the computer industry. It was the fall of 1995, a time we who are obsessed with temperature will always recall. It was the time when our friends in the personal computer (PC) industry ran into The Wall and bounced off, little cartoon stars orbiting their skulls.

"Look, here's another one at 90-MHz," Kamal said, pointing to a large ad in the newspaper. It was for a discount store called "Dungeon o' Value." They sold things like picnic tables, prams, and even PCs. Kamal chuckled cruelly.

Clement looked at the ad, which showed a kid using a PC to do his homework, and rolled his eyes. "But that's a desktop model! Anybody can make a 90-MHz desktop!"

Clement had just given a presentation about his thermal analysis of a laptop version of a 90-MHz PC. It showed that no matter what

Hot Air Rises and Heat Sinks

kind of heat sinks and vent patterns he tried, nothing could keep that microprocessor chip cool enough. The laptop case was too small for a fan, and besides, the fan would just draw down the battery too fast. Desktop PCs had at least two fans, one fan that blew directly on the heat sink on top of the microprocessor, and one that cooled the power supply and everything else. He had been forced to conclude that the only way to keep the laptop cool enough was to reduce the clock speed (and so the power dissipation) of the microprocessor.

It was a good presentation of a thorough thermal analysis. But still Clement had to bear our laughter. He had hit The Wall, and, as yet, we had not.

"Hey, I'm sorry, but if you want to carry the PC around with you, you have to make some compromises," Clement said. "So it runs a little slower. How fast can you play golf on the PC anyway?"

I looked at the resignation in his eyes, and my laughter turned to chill. I had not yet faced The Wall — the practical limit of natural convection. Laptop makers were just then hitting it, and TeleLeap, I was beginning to realize, was not very far behind.

Invest in Heat, It Always Goes Up

Clement had included the graph in Figure 10-1 in his presentation, to give us an idea of the challenge he had faced in trying to cool the latest microprocessor in a laptop case. It shows the historical trend of power dissipation of the microprocessors used in PCs over time.

It's easy to see that the power dissipated by these chips is going up year by year. The power scale is logarithmic, which means that power is not going up steadily, it is going up exponentially. This trend is happening for a couple of reasons:

- Clock frequency is going up. Video games just don't play fast enough at 33 MHz, especially if you want more and more three-dimensional graphics and color. The power of a logic gate goes up with the switching frequency.
- Silicon feature sizes (transistors, diodes, etc.) are also getting smaller and smaller, which allows the chip designers to pack more and more gates into the same area. Watts per square inch of die area is growing even faster than the watts per chip.

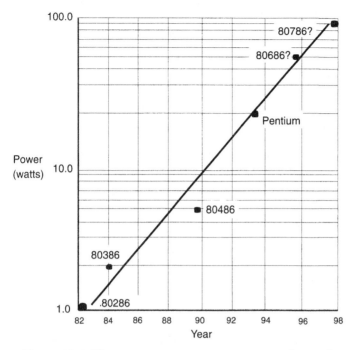

Figure 10-1 Microprocessor power grows even faster than sales

On the other hand, you have the historical trend in the capabilities of natural convection in air (Figure 10-2).

That is the problem with natural convection. Mother Nature has not been motivated by competitive forces to improve on any of her processes. Of course there are many tricks that we use to get as much as we can out of natural convection, such as making sure air vents are unrestricted, and increasing the surface area exposed to the air by adding heat sinks.

But if you put the two trends together in your head, you can see that power is increasing, size is shrinking, and eventually chip technology is going to reach the point where natural convection just isn't going to be good enough anymore. Not only is each chip getting hotter, but the trend in product size is to place more chips on each board, to shrink boards, and to pack them closer together. I began to think that The Wall may be a lot closer to TeleLeap than I thought.

Hot Air Rises and Heat Sinks ─────────────────────────

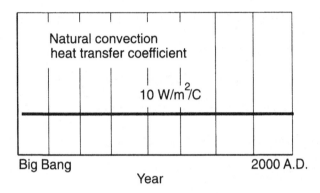

Figure 10-2 Long-term trend of average natural convection in air through history

But TeleLeap Doesn't Make PCs

"You guys are lucky you don't make PCs," Clement said, waving his mug of ale at us all. I wondered just how lucky we were.

Back home on the Silicon Tollway, I did some guesstimating by looking at the historical trends of TeleLeap product designs. In the power dissipation race, TeleLeap was probably only one or two years behind the PC industry. We had been using 4-watt chips for a few years already, and more than a couple of Coke cans had been recycled for each of their heat sinks. More and more of the TeleLeap boards I was seeing looked to me like computers — they had one or more microprocessors, RAM, flash memory, and so on. And given the life-long love affair that TeleLeap marketing managers have had with natural convection, I was starting to see "The Wall" written all over my future.

What scared me (or as my boss would say, presented me with a challenging opportunity) about this trend was that the thermal problems for our products were actually worse than for PCs. When was the last time you saw a PC designed to work at 50°C? In the telecom market, 50°C is just for openers.

TeleLeap sells to telephone central offices, not the relatively benign environment of some accountant's air-conditioned desk. PCs are expected to break down frequently, and even when they don't, to be

Hitting the Wall

obsolete junk in three years. Our systems are expected to work 24 hours a day, handling revenue-generating phone traffic without glitches, reliably, for years on end. The maximum air temperature we have to work in is higher, and our chips have to be cooler so that they will live longer. The only challenge the PC folks face that is harder than ours is stuffing their circuits and disk drives and batteries into a notebook-size package. But our board density is not that far behind. Then we throw in other high-power gizmos like laser transmitters.

When would TeleLeap hit The Wall? I put my guesstimate of the historical trend of power density in Crosser boards on a computer, and then extrapolated that trend into the future. Take a look at Figure 10-3.

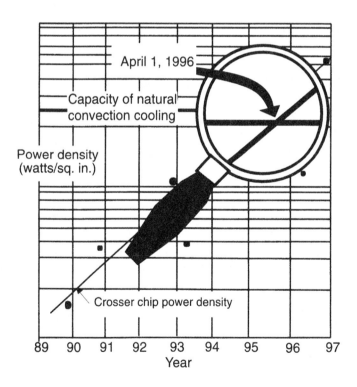

Figure 10-3 This is where air cooling runs out of gas

Hot Air Rises and Heat Sinks ─────────────────────

According to this chart, on April 1, 1996, a design team at TeleLeap would come up with a board with so many watts per square inch that natural convection just won't be able to cool it.

February 1996 — The Wall

I was ready when Herbie and his manager, Wayne, came to see me late one night after most of the staff had driven off home through the slush of late winter. They very quietly asked me, just hypothetically, what it would take to cool a board that had 100 watts instead of the typical 10 watts. I drew for them two pictures on my whiteboard. They could take their choice. One was of a heat sink bigger than the whole circuit board, with fins sticking out of the top, bottom, and front of the shelf. The other was a picture of a Radially Symmetrical Bladed Rotational Electromechanical Gas Acceleration System (RSBREGAS). They loved the name, but it looked suspiciously like a fan. They went away sad, but properly warned of the things to come.

A Temporary Retreat from The Wall

PC makers didn't just give up in 1995. They found ways to reduce power dissipation in microprocessors while still increasing the clock speed. The main thing they did was reduce the voltage from 5V to 3.3V, and in some chips, down to 2V. That knocked the power dissipation down by as much as 90% for some chips. Clement actually sent me a magazine ad that showed his 120-MHz laptop.

So thanks to that improvement, we have all slid back down the curve away from The Wall a little. Maybe we have bought another year or two. I sighed in relief, but didn't relax totally. Then I clipped a page from a catalog showing a 200-MHz desktop PC and dropped it into the mail to my colleague.

KEEPING A COOL HEAD

CHAPTER 11

> *A 25-CFM fan doesn't give 25 CFM of air, and Herbie nearly loses his head over it. Guidance for figuring out fan flow rates is given on the back of a napkin.* **Lesson**: *The fan performance curve.*

I hadn't heard from Herbie for months, and then, as welcome as a tax audit, he dropped by my office all out of breath. He'd been working in Sedona, Arizona, on one of those joint venture projects that TeleLeap likes to hook up with. It was a little high-tech start-up with an obscure new technology called Teslatronics. Instead of using wire or fiber optics, the company has a machine that allows users to hook up to the communications network telepathically. The project code name was HeadPhones. Herbie said the hardware works fairly well, at least as a prototype, but marketing is a problem. It seems that in the test runs, everybody wants to send their thoughts, but nobody wants to listen to anybody else's.

Herbie said, "We just got the last issue of your thermal newsletter about how great fans are, so we definitely want to have fans in our main system cabinet. You have 15 minutes before the limo comes to take me to the airport to tell me all about fans."

I put my feet up on my desk to get into consulting mode. "What's in the cabinet that makes you think you need fans?"

Hot Air Rises and Heat Sinks

Herbie sketched furiously on the white board. "Here are some network cards, this is the CPU, this box has the main power supply, and over here is the HBU — that part is probably the most sensitive to high temperature."

"What's the HBU? Does it have electrolytic caps?"

"Uh, not really. HBU is the Human Brain Unit. In the prototype — "

"HUMAN BRAIN UNIT?" I said, my feet plopping to the floor.

Herbie explained, "It's the transceiver link to the customer. Right now the HBU is a guy in a shielded cabinet with a lot of wires taped to his head. He receives the telepathic signals from the remote unit and downloads them to the network cards. Eventually we'll downsize it to just the essential component — a capsule containing the left hemisphere of the brain."

"You might run into a snag or two with that part of the project."

Herbie went on, "We don't have to worry about the HBU until Phase 2. What I'm most worried about is this box with the power supply. We buy it off the shelf from PowSupCo. Their environmental spec says it will work at 50°C at full load if there is 25 CFM of air. I got a sample fan from BloHard Industries that's rated at 25 CFM. But when I hooked it up to the power supply box, hardly any air came out the vents. What's wrong?"

I pulled out my dog-eared BloHard catalog and opened to the page with Herbie's fan. "First off, do you know what a CFM is?" I asked.

He made a face as if he had stepped on something squishy with his bare feet. "I assumed it was one of those metric units like a centifemiron."

"Close. It means cubic feet per minute. It is the volumetric flow rate of the fan. Every minute a 25 cubic foot chunk of air comes flopping out of it."

"Big deal. It's how much air comes out."

"You also have to read the fine print — 25 CFM is the *maximum* flow rate for this fan. That's how much air would pass through it if you hung it from a string in the middle of a large empty room. As soon as you attach it to something like a box or a grill the flow rate goes down, because those things resist flow. How much it goes down is shown in this chart here" (Figure 11-1).

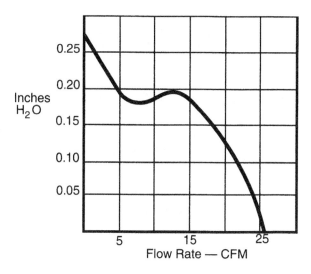

Figure 11-1 Performance curve of Herbie's "25 CFM" axial fan

Herbie bit his lip. "I saw this, but it didn't make any sense. CFM I think I get. But what the heck does H_2O have to do with fans?"

I said, "Inches of water is an arcane but sometimes useful way of measuring air pressure. They could have put it in pounds per square inch, like tire pressure, or Newtons, or atmospheres, I suppose. The chart tells you that it takes air pressure to make air move against a resistance. The higher the pressure caused by the resistance, the lower the airflow. See, here at zero resistance you get 25 CFM."

"OK, that sort of makes sense," he said. "So how much is an inch of water worth?"

"This'll give you an idea. You're a weather junkie, right? So you are already used to the idea of air pressure in inches of mercury. The barometer is at 30.07 inches, or whatever. The way to make a barometer is to take a long glass tube that is closed off at one end and fill it up with mercury. Then turn it upside down and dunk the open end into a pot of mercury (check out Figure 11-2). The mercury flows down a little, creating a vacuum at the closed end of the tube. The air pressure outside balances the weight of the column of mercury and keeps it from flowing out. Then glue a big ruler on the side of the

Hot Air Rises and Heat Sinks

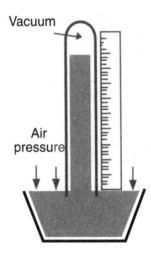

Figure 11-2 The barometer: How pressure got to be in inches

tube. When the air pressure changes, the mercury moves up or down in the tube. Air pressure is measured by comparing the height of the mercury in the tube with the ruler, which at sea level is normally around 30 inches of mercury."

Herbie bought the weather angle. "So inches of water is like barometric pressure."

"Not exactly, but we're getting there. You can do the same thing with water. But there is a good reason why nobody does it, and that's because mercury is so much more dense than water. A mercury barometer is already a pain in the neck to have around because it is nearly three feet long. A water barometer would be about 34 feet tall."

"Inches, not feet," Herbie said. "You said feet."

"No, I meant to say feet. You might just fit a water barometer inside the arboretum of the R&D building, but you'd have to stand on a chair on the third floor landing to read it."

Herbie squinted at the fan chart closely. "But the fan chart is in inches of water, not feet. If you're right, then this fan doesn't put out very much air pressure."

"That's what I'm getting at. An inch of water is not very much pressure at all. Let's see... if 34 feet of water is about 1 atmosphere, and this fan puts out about 0.1 inches of water, that's about ... 0.0002 atmospheres. It doesn't take a lot of flow resistance to stop that kind of air pressure."

"So there's my problem," Herbie said, scratching his head. "My fan is too wimpy for the job. But how, Uncle Thermal, do I figure out what size fan to use?"

"There is the thermal design section of the corporate *Good Design Guidelines*," I rattled off. "It tells all about fan curves and system curves. You could read it, or I could tell you — for a price." So over a free coffee I sketched on a napkin the following (Figure 11-3):

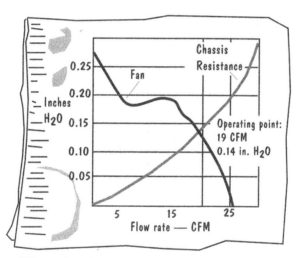

Figure 11-3 Where the chassis resistance curve (not the coffee stain) crosses the fan curve is the operating point

"Just like your fan has a flow versus pressure curve, your cabinet has a curve something like this," I said. "It goes in the opposite direction of the fan curve. The more flow you want, the more pressure it takes. If you knew what the curve was for your cabinet, you could draw it on the fan chart, and where the two lines cross — Bingo — that would be the operating point, and it would tell you the actual flow rate you would get."

Hot Air Rises and Heat Sinks

"So where can I look up this cabinet curve? Have you got a handbook or cheat-sheet someplace?" Herbie asked.

"Nope. It all depends on your cabinet, the size and shape and arrangement of the stuff in it, cabling, hole patterns in the grills, turns in the flow direction — all kinds of details. Even after you build your box, it's almost impossible to measure without a fancy wind tunnel. And because I can see out the window that your limo is waiting at the curb, I'd better just give you the Rule of Thumb to keep you going for now: Get a bigger fan than you think you need. Pick one with a rated flow that is about twice as much as the flow rate you want. Because you need at least 25-CFM, you'll have to find a fan that is rated at about 50 CFM. Then put it together and test it."

Herbie pocketed the napkin, took a last slurp of his coffee, and made a dash for the limo. I'm sure I'll see that graph in a report some day, coffee stain and all. But at least now he knows that he can't get 25 CFM out of a 25-CFM fan. Later I'll call him and tell him that the thermal spec for the power supply is too vague to be useful, and that he might need to account for altitude changes in his testing, and that, for the sake of the person in the HBU, I hope he's planning some kind of redundancy in case the fan breaks.

Temperamental Prototypes

CHAPTER 12

> *Cooling an electronic component is different from cooling a power supply is different from cooling a human being. Setting thermal design goals for a project is more than just filling out forms.* **Lesson**: *Operating temperature limits.*

It's amazing how much emotion can come through in a simple e-mail. I almost felt the hug behind the "Happy Birthday" my mother-in-law e-mailed from Boulder. I definitely sensed the anger transmitted with the following message from Herbie one morning:

> Tony,
>   How come you Quality Simulation guys never consult with us design engineers when you revise the flippin' *G-D Guidelines* [*Good Design Guidelines*]? Part of the deal we made with these guys in Sedona on the HeadPhones project was that they would fill out all *G-D* checklists, just like we do back at TeleLeap. Only they aren't doing the checklists just like we do, THEY TAKE THEM SERIOUSLY.

Hot Air Rises and Heat Sinks

> Since I did such a great job with the fans on the power supply last month, they assigned me the job of filling out the Thermal Design Checklist. I figured that would be a snap, since we've been filling them out for over a year now in my old department. But you guys sent out a revision to the *G-D* a couple of months ago. I ran smack into a road block right away on Question 2. It says, "Have Thermal Design Goals been identified in written form?"
>
> First off, what the heck is a Thermal Design Goal? I can't find anything about that in the Project Mission Statement. Since when do I have to put them in written form? I never had to before. I was going to check the "Not Applicable" box like I usually do for questions that I don't understand. But the Sedona folks thought that if the *G-D* wanted something in writing, it must be pretty important, so they want to comply with it.
>
> Respond immediately with a Letter of Exemption for me, OK? E-mail is OK, but if you could follow it up with something on company letterhead that would be even better.
>
> the herbster

I sent off the following response and went to lunch.

> Hello Herbie,
> The recent *Good Design Guidelines* revision did not have any changes in

Tempera-Mental Prototypes

the Thermal Design Guidelines or
checklists. Take a peek at the revision box on the top of any page and
note that it is still at Rev A. The
checklist has not changed since Day
One. The WRITTEN list of Thermal
Design Goals has been required for
every project since Day One also.
Don't panic, but there are two other
written requirements in that checklist
that you haven't got to yet: the
Thermal Test Plan and the Thermal Test
Report.
   You don't get an exemption letter.
Thermal Design Goals is a fancy way of
asking the question, "How hot do I
want my product to get?" If you don't
know the answer to that question, how
can you design the cooling system? And
how do you know if it works or not after you build it?

                              Good luck.
                        Mr. Good-Design

After a quick lunch, I came back to my office to find the computer practically jumping up and down on the desk. Herbie had responded:

Hey Heatsink-breath,
   We found your letter of exemption
not acceptable. I can't check the Not
Applicable box until you give me
written permission. These Sedona guys
are really getting picky. They think
that someone at TeleLeap is actually
going to review how we answer these
checklists after we hand them in at

Hot Air Rises and Heat Sinks

> AFLR (Absolutely Final Last Review).
> Nobody ever wrote a written Thermal Design Goals before, and you and I both know it. I'm pretty sure nobody ever wrote a Thermal Test Plan either. So get with reality. I need to finish this before I can go home for Christmas.
> Herbenstein

I tried calling Herbie, but all I got was a generic voice mail greeting, so I e-mailed again.

> Dear Herb,
> I'm not trying to throw more hurdles in your way. I want to help you jump over the ones that are there, whether they are visible to you now or not. There is no such thing as an exemption, so I can't give you one. Just write the Thermal Design Goals already. It isn't that hard.
> Let me give you an example of how such a huge document might appear:
>
> **THERMAL DESIGN GOALS**
>
> **Project: HeadPhones Power Supply**
> General electronic component thermal limits:
> The maximum allowable temperatures for electronic components when the system is operating at its worst-case environmental conditions are given in Table 2.5.1.1-2 in the Thermal Design Section of the *Good Design Guidelines*.
> Special temperature limits not covered by the table listed above:

1. The heat sink of the power supply brick 79.XXXX shall not exceed 85°C during any operating condition. (Herb, this is just an example of something you might want to put in, not something that is always required. You need to look at the list of components you might be using and decide if they have any special temperature limits. The *G-D Guidelines* only cover reliability. You have to figure out how hot your components can get and still do the job you want them to. Like that one line driver you told me about, where the leading edge of the signal started to round off more and more as the chip got hotter.)
2. Surface temperature of Emergency OFF handle shall not exceed 70°C, to prevent injury to the operator. (Again, just an example. You might have controls, or handles that people have to grab onto, and you don't want them to get burnt.)
3. Other, non-electronic temperature limits that you might have to worry about, depending on what you've got in that secret project of yours. Printed circuit boards typically should not get over 105°C because the epoxy starts to get mushy, and you can't maintain the spacing between traces unless you use a higher temperature-rated material. You might need to worry about the

Hot Air Rises and Heat Sinks

temperature ratings on the insulation of your cables and other plastic parts. If they get too hot, they can change properties, lose their fire-retardant chemicals, or even melt. If you have motors, then you don't want to bake off the lubricants in the bearings. I can't make you a list of these things that works for every type of product. That's why I want YOU to make the list as you design it.

The End.

That's all there is to it. The first part refers to a table in the *G-D*. That table says, and I quote, that the junction temperature of memory devices must not exceed 125°C. The second part is for anything else in your design that isn't covered by the table.

If you want, just copy this e-mail, then edit it for your specific project and print it out. Review it with the Sedona design team to see if it's right and complete, and then you're done. Check the box in the *G-D* checklist and keep going.

It wasn't until after my afternoon coffee that I got another message.

Uncle Thermal:
It always looks easy to you guys up in the corporate ivory tower. Just fill out another form, another check-

list, another little one page report that takes three days to write. Oh, no, to save yourself from writing one little exemption note, a measly e-mail for Pete's sake, you practically hold a gun to my head and force me to write something NOBODY else in the whole company ever had to write before.

OK. I give up. I'll write the durn thing. I copied the stuff out of your message about Table 2.5.1.1-2 to my document. But I don't think there are any power bricks or emergency OFF things, so I deleted that, but there is one thing I'm not sure how to put in there. And that's the whole reason I think we still should get an exemption. Because none of the *G-D* talks about how to deal with this kind of component.

I'm talking about the HBU, you know, the Human Brain Unit. In Phase I it consists of this guy sitting in a recliner inside a sheet metal box with a bunch of wires taped to his noggin. He gets the telepathic signals from the Remote Unit, and — well, I better not say too much. It's supposed to be secret.

So what do I put for a temperature limit for this guy? 98.6°F? If he gets too hot, won't he just sweat or something? Shouldn't this be "Not Applicable"?

Herbarama

Hot Air Rises and Heat Sinks

Finally I knew what all the emotion was about. Herbie wanted to skip the question not because it was irrelevant, but because it was crucial.

> Herb,
> This is not "Not Applicable." It is 100% applicable, and 100% necessary for you to answer this question. That's the whole point of making you go through the checklist. After AFLR, I won't care how you checked off the boxes on a piece of paper, and nobody else will either. At that point you will have built a working version of your design SOMEHOW, which means you will know how hot a guy can get before he stops being telepathic. Maybe by then you will have discovered it by trial and error. What I want you to do is to figure out that temperature limit now, in the early part of your design, and avoid the trial and error process.
> Just to get you started, I have a few estimates about the operating temperature limits of human beings. These are all ambient or room temperatures, not body temperatures. The absolute minimum temperature a person can stand for just a few minutes without protective gear is about 5°C. The long-term minimum is about 15°C, and the long-term maximum is about 28°C. The absolute maximum is no more than 50°C. After a few minutes, the HBU would probably pass out, and after a few

> hours he might even die. The optimum average air temperature is probably around 21°C, but maybe he gets more telepathic at higher ambient. You will have to do a few experiments to find out the real limits. Note that all of this is just a guess. And I think humidity probably has a big effect on it, too.
> Let me know how it turns out.

Just before I went home, I got this reply:

> My Hero,
> Got your last message and have used it to write up my Thermal Design Goals. Per your spec, I put in the HBU limits as +5 to +50°C. Thanks for all your help. See you after Xmas.
>
> Herb

That's why Herbie has gone so far in his career. He knew how to take very little information and run with it. I had to help him a lot more with the details later. But he had made a great thermal leap forward. He had written Thermal Design Goals. A first for a design engineer at TeleLeap.

For years, thermal design at TeleLeap had been done backwards. The engineers would design and build a new product, then test to see whether it worked in the environments that the customer might want. Lots of times they were lucky and no big problems were found, so they figured they had a good design process — it was usually successful. But you never learn anything from success.

A few products did have thermal problems. Sometimes they weren't discovered until the product was out in the field. Then came the Band-Aids and the heroic and expensive redesign efforts. After a few of those, the Quality Simulation Department came up with the

Hot Air Rises and Heat Sinks

Good Design Guidelines, called the "G-D" Guidelines for more than obvious reasons. The checklists in this document have forced the old design process to turn around and inside out. So designers like Herbie have to think about actual component temperature limits *before* they finish their prototypes.

I hate checklists and procedures, too. But until thermal design is brought out of the land of myth and mistake into the land of engineering common sense, the HBUs of designers will have to heat up doing this extra paperwork.

Misdata

CHAPTER 13

> *Component data books are chock-full of specs that are only valid when they are not important, just like my temperature-telling watch, which is only accurate when it isn't too hot or too cold outside.* **Lesson**: *Temperature limits in terms of air temperature aren't very useful*

In the old days, before they came out on CD-ROMs or were distributed over the Internet, everybody's work area was full of them, those overstuffed paperback books (and I'm not talking about the Harlequin romances hidden behind the Tupperware sandwich box). Data books. I think they should be called "misdata books." Misdata, as in misinformation.

And just because they come in a more convenient format doesn't mean the information in them has gotten any better. Motorola and National and the other good quality component vendors in the world were not trying to deceive anybody, nor did they bungle and put a lot of misprints or incorrect numbers in those billions of onion-skin-thick pages. They were just following industry standards. But I find it very annoying that some of the numbers in those data books are only correct in situations I don't care about.

To be specific, I want to complain about θ_{J-A}. The data book says θ_{J-A} is the thermal resistance of a component between the junction (**J**) and the air (**A**) (Figure 13-1). The idea is that if you know the

Hot Air Rises and Heat Sinks

power (**Q**) dissipated by the component, and you know the air temperature, you can figure out the junction temperature from a simple equation (don't stop reading, there isn't that much math here):

$$T_J = T_A + Q\,(\theta_{J\text{-}A}) \qquad (13\text{-}1)$$

For example, from the Texas Instruments data book *Advanced CMOS Logic*, a 16-pin DIP has a value of 110°C/W for $\theta_{J\text{-}A}$ in still air. So if the power is 0.5 watts, then the junction temperature will be 55°C hotter than the air. Simple enough. And like anything that simple, it's probably wrong.

It reminds me of my really cool Chronik World Time and Temperature Wrist Chronometer. Because temperature is my middle name, and the alarm–silence button on my old watch broke, my wife decided this temperature-measuring watch would be perfect for me, and out of the kindness of her heart, she told me to buy it for myself. Besides all the beeps and lights every other digital watch has, this thing tells the time *and* the temperature in cities all around the world. It doesn't really know the temperatures in those cities, but it has a table where it can look up the average high and low temperatures for the current month, which is probably good enough for anybody who isn't in one of those

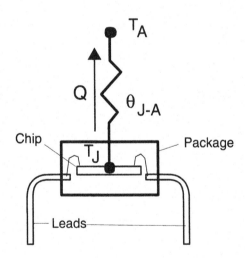

Figure 13-1 This simple model is the beginning of misdata

cities at the time. I can look at my watch and say it's –10°C in Moscow, and it's accurate enough no matter what the real temperature is, because I don't really care. The only time I might care is if I were *in* Moscow. For that situation, the watch does something else — it has a built-in temperature sensor that actually measures the local air temperature and gives a continuous readout.

This is where my watch story starts to connect with my tirade against θ_{J-A}. Because the thermal sensor is built into the body of the watch, the only thing the sensor can measure is the temperature of the watch itself. Chronik designers are not morons. They know that nobody wants to buy a watch that tells you the temperature of the watch. It's pretty obvious that the temperature of the watch is affected by the heat from my arm and the heat generated inside the watch, as well as the nearby air temperature. The little data book that comes with the watch, printed in umpteen languages, explains that, yes indeed, the heat from your arm does affect the temperature measurement, but the computer inside the watch "compensates for body temperature to give you an accurate air temperature." People ask me about that all the time. "Doesn't the heat from your arm screw up the temperature?"

"It *compensates*," I say. I don't really know how it compensates. If you wanted to make a really accurate air thermometer for your wrist, it would have to be complicated. This Chronik already tells time and beeps to tell you when to eat lunch and go to the bathroom, so for $59 I don't think they could put a whole lot of extra sensors inside that could compensate properly. My guess is that the little computer inside takes the watch temperature measurement and does something simple to it, such as subtracting 7°C or multiplying it by a fudge factor, like 0.8.

The reason I suspect this is that the temperature reading is actually pretty good when I am indoors at comfortable room temperatures. I have compared my watch readings with expensive thermocouple data loggers and against mercury thermometers. When the air temperature is around 24°C (75°F), the Chronik agrees with them within about 1°C, which is pretty darn good. But when I go outside and I'm sweating in the Colorado desert or shoveling snow before dawn, I want to know the air temperature so I can whine about it to my wife later. In those situations, the watch temperature reading is always way off.

In sunshine, the black body of the watch gets much hotter than the air, or even my arm, and gives a reading much too high, say, 49°C, the highest reading on the watch. It definitely does not compensate for solar radiation effects. And in the cold, the watch always reads too hot again. I'll be standing knee-deep in snow, staring at the alcohol thermometer on the side of my garage that reads –2°C, and my watch reads 17°C. The bigger the difference between my arm temperature and the air temperature, the bigger the error in the reading.

Now back to $\theta_{J\text{-}A}$. It is a tool that some data books want you to use to estimate junction temperature. But there are a couple of flaws in it that limit its usefulness from the very beginning. The first one is the assumption that you know T_A, the "ambient" temperature. There isn't a good definition, or even a hint, of where you would measure this temperature. It can't be the room temperature outside the equipment, because obviously the component temperature depends a lot on the nearby air temperature, which could vary quite a bit inside the chassis or shelf. If "ambient" means the air near a component, then we need to find out how near and whether it is upstream, downstream, or centered on the part. In fact, there is no possible way to define this local "ambient." Figure 13-2 shows the way it is interpreted by semiconductor manufacturers.

A component is soldered by itself to a test board, and then it is put into a test chamber. The test chamber is very large, compared with the size of the single component, so it is possible to control the air temperature inside it. The temperature of that great bulk of air is what they call T_A. It is under this totally unrealistic condition that $\theta_{J\text{-}A}$ is measured. They power up the component and measure the junction temperature, the chamber air temperature, and the power dissipation. With those three numbers they can calculate a value of $\theta_{J\text{-}A}$.

Compare that with what happens when you stick that component on a real circuit board with lots of other components in a system with realistic airflow. Figure 13-3 shows some realistic isotherm lines in the air near a circuit board.

Misdata

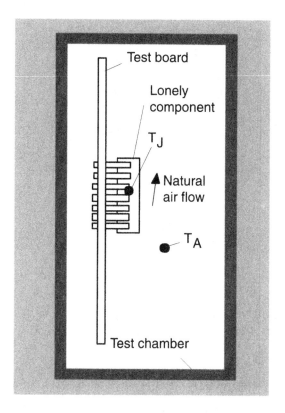

Figure 13-2 What "ambient" temperature means to a component vendor

What would be the right T_A for a component located somewhere in the middle of a board, where the air temperature can vary 60°C in less than an inch?

The second flaw is the assumption of "still air." In the manufacturer's test, there is no fan pushing the air. In our real circuit board situation, there is no fan either, but you could hardly call the air "still." Because of all the heat from other components, the air heats up and moves because of its buoyancy. The air velocity next to our component is a lot higher on the real circuit board than when it is all alone on that test board. The higher the air velocity, the better the heat transfer, which means the junction could actually be cooler than θ_{J-A} predicts.

There are other problems with θ_{J-A}, such as that it depends on how much copper is in the test board, and also that the test board

Hot Air Rises and Heat Sinks

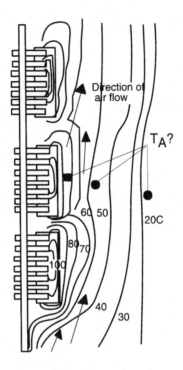

Figure 13-3 A realistic air temperature profile

itself is not standard from one manufacturer to another, but we don't need to get into that. By now I've shown that θ_{J-A} is nothing but a lovely example of misdata. It would only be useful the next time you were designing a board with only one component.

So what should you use instead for estimating junction temperature? After all, I have berated Herbie more than once for not basing the design of his cooling system on junction temperature. There is another number you can sometimes get from data books, called θ_{J-C}, which is the thermal resistance between the junction and the outside of the component package, called T_C (case temperature). It has some serious problems of its own, which I'll tell you all about when you're a little older, but it does have one advantage: you know exactly (more or less) where to measure T_C.

CHAPTER 14

Pessimism: A Tool of Quality

> *Herbie and Vlad discover that two fans are not always cooler than one.* **Lesson**: *Fans in parallel don't always provide redundancy.*

We were having a few beers after dinner. Herbie's girlfriend, Vernita, wanted me to take one of those quizzes from her favorite magazine. "It tells if you're an optimist or a pessimist," she said.

"I already know I'm a pessimist," I said.

"OK," she said. "The rest of the article tells how you can change into being an optimist!"

"But I'm perfectly happy being a pessimist. Why would I want to change?"

Herbie eyeballed the level in his beer mug and said, "How can you say you're happy? You'd say this mug is half empty, which is bad, but we, being optimists, say it's half full, which is good, and makes us happy!"

"What's with this half-a-glass business? Fill it UP!" I grabbed the pitcher and filled up my own mug, then topped off Herbie's and Vernita's. "Now that's pessimism in action."

Herbie was buying me dinner because I was in Sedona to help with the thermal testing on the HBU (Human Brain Unit) for the HeadPhone Project. You may recall that the HBU mainly consists of a telepathis guy sealed in a box with a bunch of electronics hooked up to his brain. The dinner was Herbie's way of getting me to be the one to attach thermocouples to Vlad, the guy in the box. Vlad was already temperamental, having to shave his head every day, and I didn't think he was going to be happy with where we needed to stick the temperature probe.

Pessimism is a very useful design tool. "Expect the worst, and you will never be disappointed," is one of my mottoes. Perhaps it could better be described as Constructive Pessimism. Don't confuse me with someone who sees the world as a bad place and just whines about it. I see the world as a place where lots of things go wrong, and it's my job to expect that and look for ways to prevent it as much as possible.

It takes a pessimist to really appreciate this question from the *Good Design Guidelines* (G-D) Checklists in the part on fan-cooled systems:

Has the worst thing that will happen when (not if) the fan fails during worst-case operating conditions been identified?

Because fans fail, you have to be ready to deal with the consequences when it happens at the worst possible time. Commercially available fans have lifetimes rated somewhere between 20,000 and 100,000 hours of continuous operation. That is between two and eleven years. As a pessimist, I expect closer to two years than eleven, and even in today's crazy market, two years is not the expected lifetime of a telecom product. Not only that, but the failure curve for a fan is pretty much like a light bulb — they fail randomly in time. Some poop out right away, some after only 90 days, some in 1 year, and some after 20 or 30 years. It helps to think of fans as light bulbs: they are expendables that might need to be replaced at any time, and they ought to be easy to service.

Sometimes it is not so obvious how to answer this pessimistic checklist item. Herbie thought he had done it. But the testing we did the next day revealed a few problems with his scheme to deal with fan failure. The problems arose because an optimistic viewpoint had been adopted as the design evolved.

Pessimism: A Tool of Quality

Revision A of the HBU was a box without any box. Vlad just sat in a chair in a regular room. But that didn't work because he was getting too many stray brain waves from the environment. Revision B put Vlad and his chair in a shielded box, about like one of those 3-Photos-for-$1 booths. The box was lined with a material that thoughts could not penetrate — I think it was copies of a tabloid newspaper. Revision C appeared a few minutes later when they had to punch a few air holes in the walls for Vlad. For Revision D, power supplies and transceivers were built into the base of Vlad's chair, which added about 500 watts to the balance. When Vlad started to perspire, Revision E appeared. A single, 100-CFM exhaust fan was added to the back wall of the box.

Revision E actually worked pretty well in early thermal tests, but Herbie decided that it needed to be more reliable, as the fan was bound to fail sooner or later. The estimated reliability for the system was pretty good before the fan was added, although I don't think it was fair to put an estimated life of 72 years into the equation for Vlad. Herbie's scheme for improving the reliability of the system was simple — he added a second fan and a fan failure alarm. Tests on Revision E showed that one fan by itself cooled the box just fine, so it was obvious that two fans side by side would be even better (Figure 14-1). If one fan failed, the other could keep the box cool until a service technician could replace the dead one. The fan alarm would alert the service tech in plenty of time so that it would be very unlikely the spare fan would also fail during the repair period, so the whole system would stay up and running.

We spent half the next day attaching thermocouples to Revision F. After much discussion, Vlad allowed us to tape the body temperature probe in his armpit. We sealed Vlad into the box, turned on the electronics, started up all of the Telepathic-to-TTL Loopback test sets, and watched the temperatures on our data logger. After an hour, temperatures stabilized, and Herbie grinned at me. "Looks like everything is way lower than the temperature limits from my Thermal Design Goals," he said. He waved the document in front of me proudly.

"OK," I said, "now unplug one of the fans."

Hot Air Rises and Heat Sinks

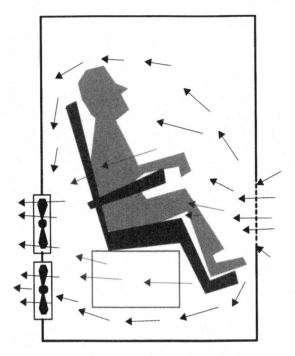

Figure 14-1 With both fans working, Vlad is cool

"But we don't need to," Herbie said. "We tested with just one fan last fall and everything was fine. If I unplug a fan, sure the temperatures will go up some, but we already know one fan can do the job by itself."

"Let's see what happens." I had a feeling that it might be important to test the fan redundancy feature. Herbie grumbled, then crawled behind the box to get at the fans.

"Son of a — ! Looks like we'll need a change here," came Herbie's muffled voice. "There's no way to unplug the fan. The wires come out of the fan housing and go inside the box to the power supply. I'd have to go in there to disconnect it."

To keep the test going he snipped the power wire to the fan. The temperature curves on the data logger screen started to creep up a little. "See, I told you," Herbie said. A red LED on the box lit up and a message appeared on the system console: FAN ALARM.

Pessimism: A Tool of Quality

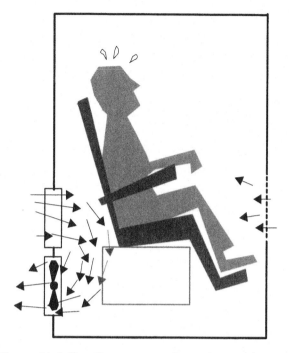

Figure 14-2 Two fans are not always better than one

After a few minutes, though, one power supply got much hotter than the other, and its thermal safety switch shut it down. This immediately doubled the load on the remaining power supply, and its temperatures started to go up fast. It was approaching thermal shut down, too, when I noticed Vlad's armpit sensor suddenly jump about 5°C, and then the sound of Velcro ripping open. Vlad burst out of the box, furiously slapping at his rear end. The power supply heat sink under Vlad's chair had nearly started his beloved polyester pants on fire.

After a little first aid, we diagnosed the problem. When Herbie added the second fan to the back of the box, he also added a second HOLE, of course, for that fan to draw air through (Figure 14-2). That didn't matter until one of the fans quit. Then that extra hole stopped being a fan outlet and became an air LEAK. The air, instead of flowing through the tiny inlet ports around Vlad and through all of the guts of the power supplies and transceivers, would rather flow in

Hot Air Rises and Heat Sinks

through the dead fan and right out again through the good fan. There was still some airflow near the working fan, which is why one power supply was cooled more than the other, but essentially there just wasn't enough air moving through the box to keep the system working.

Vlad was not seriously hurt, but there was a lot of hard staring at the big meeting that took place that afternoon. "And why was the heat sink right under Vlad's, uh, seat?" a project manager asked.

"The *G-D Guidelines* recommend bolting hot components to the chassis. We got the power supply to work by using the chair as a heat sink. It only gets super hot if one supply is carrying the full load — and the fan dies, which is a double fault condition. Or at least we used to think so," Herbie explained.

Vlad was not happy and threatened to leave the project and take the other seven telepaths with him. The future of telepathic communication was in jeopardy. A raft of design changes was proposed: adding more fans, spreading them out, inserting baffles, opening up inlet air ports, moving the power supplies. Suddenly everyone in the room knew exactly how the box should have been designed in the first place. Action items were assigned and the meeting broke up in time for a late dinner.

It was a very quiet dinner. Near the end, Herbie pulled out a crumpled page from the *G-D Checklists* and read out loud, " 'Has the worst thing that will happen when (not if) the fan fails during worst case operating conditions been identified?' I thought so, but I didn't think a fan failure could lead to the whole project going down the tubes."

I stared at my not-quite-full glass of beer. "Do you want to take that Pessimist–Optimist test again?" I said.

Blowin' in the Wind

CHAPTER 15

Junk science and misconceptions in heat transfer. Where does all this baloney come from? Start with the TV weather and the "windchill factor." **Lesson**: *Forced convection heat transfer equation.*

Winter has my least favorite weather, and not just because from October until June, Illinois is one big Slushfest. It's the wintertime weather reports, especially the windchill factor. When I hear Tommy Temperate on TV talk about the windchill, I cringe the way an English teacher does when you say "irregardless" when you mean to say "hopefully." Why does it bug me so much?

For one thing, the windchill factor is not a factor. An example of a factor is the number 7 for "dog years." If a dog is 3 years old, you multiply that by the "dog year factor" to get the equivalent number of "people years" of 21. On the weather, they say it's 12°, with a windchill factor of –10°. Minus 10° is not the *factor*, it's the *result* you get after you apply the windchill correction factor to the actual temperature of 12°. They don't ever tell you the factor, they only give you the "corrected" temperature.

It also bugs me because it's not based on Heat Transfer Theory, it's based on TV ratings. The lower the temperature, the more people

Hot Air Rises and Heat Sinks

watch the weather. So the windchill factor was invented to make the weather reports more spectacular rather than useful.

If you hearken back to the story in Chapter 3 of how circulating fans inside an environmental chamber lead to erroneous temperature tests, you will see that I am showing you another face of a familiar idea. The whole idea of windchill is that when the wind is blowing, it *feels* colder. Our senses are not lying to us. But the phenomenon that our nerve endings are reporting is not temperature, but the rate of heat loss passing through the skin into the air. When the wind blows, the air temperature doesn't change, but the rate of heat loss does change. The Heat Transfer expression for this is

$$Q = h A (T_{skin} - T_{air}) \qquad (15\text{-}1)$$

where

Q = how fast heat is lost from your skin
A = the surface area of skin that is exposed to the air
$\mathbf{T_{skin}}$ = the temperature of the skin
$\mathbf{T_{air}}$ = the temperature of the air
h = the convective coefficient, or fudge factor, that is directly related to the speed of the air passing over your skin. The faster the wind blows, the bigger **h** gets. The bigger **h** is, the greater the heat loss is at any particular air temperature. So when the wind blows harder, you feel colder (Figure 15-1). The equation was developed by guys like Fourier and Isaac Newton, and has been in use a lot longer than the "science" of weather reporting, so you can put some trust in it.

There is another feature not so obvious in this equation. Heat loss depends on the temperature difference between your skin and the air. If your skin cools down to the same temperature as the air, the term $\mathbf{T_{skin}} - \mathbf{T_{air}}$ becomes zero, and the rate of heat loss also goes to zero. That means that no matter how hard the wind blows, the minimum temperature your skin will ever become is the air temperature.

Compare this to the intuitive idea you get from the weather report by taking this little pop quiz.

Blowin' in the Wind

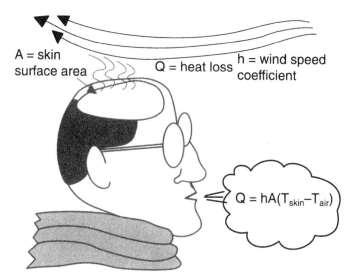

Figure 15-1 A is the "exposed" surface area

Question 1. It's January, and on the Weather Channel you hear it's going to be overcast and 38°F all night long. The wind is blowing something fierce, and WC reports the windchill factor will be 10°F all night. As you climb into bed, you remember the dog's water dish is out on the back step (and that water freezes at 32°F). You don't want the water to freeze and crack Fido's autographed Rin Tin Tin dish. Do you need to bring it in the house overnight?

Question 2. Crossing guards, TV weatherfolk, moms, and other pseudo-officials always warn you in winter to wear a hat and scarf outside because you lose 50% of your heat through your head and neck. Does that mean that when you are standing at the bus stop in your coat and boots and no hat, you would be just as comfortable standing there totally naked, except for a hat and scarf?

Answer to Question 1. The air and water will not ever go below 38°F, if the Weather Channel prediction is actually right. That is why your dog's water won't freeze if the air temperature never goes below 32°F, even if the windchill is –20°F. (The smart guy in the back of the class is asking about radiation losses to the sky, but it is *overcast*, so I can skip all that for now.)

What does TV Weather Theory predict? TV math works something like this. After looking at the gages for average wind speed and air temperature, they look in a table or a graph to find the windchill temperature, T_{wc} (Figure 15-2):

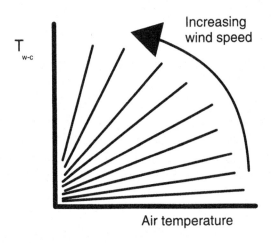

Figure 15-2 The windchill graph every TV station has somewhere

Then, to find the rate of heat loss from the skin, you'd use an equation like this:

$$Q = KA\,(T_{skin} - T_{wc}) \qquad (15\text{-}2)$$

In the "TV Heat Transfer" equation, **K** is some kind of conversion factor to make the units come out right. They've thrown out **h,** the convective or wind speed factor, and tried to make up for it by using the "corrected" temperature in place of the actual temperature. This approach leads to a misconception that the lowest temperature a body can reach is the windchill temperature, which leads to the strange sight in the winter of people putting blankets over their car engines overnight to protect them from the windchill.

Maybe I need to explain why putting the blanket on the car engine overnight is a waste of time. It seems intuitive that it should help. After all, if you wrap yourself in a blanket, it helps you stay warmer, especially if the wind is blowing hard. The blanket protects your skin from windchill. Why won't it protect your car engine from windchill?

Look at the heat transfer equation again. If you cover your skin with a blanket, you are reducing **A**, the area exposed to the air, and so the rate of heat loss is reduced. This is good, because now you can maintain your skin at its normal temperature close to 98°F without losing a lot of heat.

But the second you shut off your car, the engine stops producing heat (unlike your body), and within a few minutes, the engine temperature drops nearly to air temperature. The point of putting a blanket over the engine is to keep it from dropping all the way down to that scary-sounding windchill temperature. But the blanket won't do anything. Once the engine reaches true air temperature, it won't go down any further, whether you put a blanket on it or not. It would make a difference if the engine were left running, generating heat. But the last time Herbie did that, the blanket got caught in the alternator belt, which stalled the engine anyway.

I have another complaint. How come they don't have a wind*burn* factor? Normal body temperature under your tongue and in other unmentionable probe locations is about 99°F. Your skin is less than that, about 90°F. In summer, the air temperature often exceeds 90°F, which means that the air is hotter than your skin, and that means heat will enter your skin from the air. The equation works in either direction. The harder the wind blows, the more heat "gain" you get, which means you should feel hotter. Shouldn't there be a windburn factor to tell you to cover up and protect yourself from the hot air blasts?

Hot Air Rises and Heat Sinks

Luckily, the human body gets rid of more heat by sweating and breathing than it gains by convection. And the faster the wind blows, the better sweat evaporates, so the windburn factor is more than canceled out by the armpit factor — unless the humidity is close to 100%, which prevents evaporation. But that is another story. If electronic components could sweat, that topic would be worth continuing.

There is something to be gained by this tirade against misleading weather reports. The heat transfer equation can be your friend. It has been mine. As Dr. Van Allen, the discoverer of the radiation belts surrounding the earth, told a reporter who had naively asked him if the Van Allen Belts had any practical purposes for people on earth, "Well, I've made a pretty good living off them so far."

This equation, besides being useful for calculating temperatures and rates of heat flow, conveys some basic notions for cooling electronics with air.

- Heat flows from hot to cold. No matter how big a fan you stick in the chassis, you can't make the heat come out of the components if the air is not cooler than the components. This might seem obvious, but Herbie's boss did once ask me how big a heat sink it would take to get the Cyclone chip down to 45°C. As the air temperature entering the shelf could be as high as 50°C, I had to tell him that the heat sink had to be big enough so that one end of it could reach his refrigerator.
- The faster the airflow, the better the heat transfer. In general, putting a fan in the system will lower the temperature of all components compared with having no fan. But don't forget about the "windburn" factor. There might be some low-power components, such as capacitors, downstream from a large power supply. If that power supply heats the air so much that it is higher than the capacitor temperature, then heat will transfer from the air into the capacitor, raising its temperature. And the faster the airflow, the hotter that cap will get.
- The more surface area exposed to the air, the better. Don't cover your components with blankets (or closely spaced daughter boards) if you want the heat to get out.

On Question 2, about comfort with a hat and scarf and no other clothes, compared with pants, coat, gloves, and no hat — I'm pretty sure the answer is that mom is right. You should wear a hat and scarf, but mostly because it is a trick question. When she says you lose 50% of your heat from your head during the winter, she means *if you already have a coat and pants and boots and gloves on.* Which means that the amount you are losing through your head is EQUAL to the amount you are also losing through your well-insulated coat and pants. Plus, it's kind of complicated, because your blood circulation will rob heat from your chest and arms and legs to keep your brain warm, which makes you lose heat even faster, so this analogy with electronic components starts to get really lame. Just wear a hat.

THERMOCOUPLES: THE SIMPLEST WAY TO MEASURE TEMPERATURE WRONG

CHAPTER 16

> *The most reliable and accurate way of measuring component temperature can also literally explode in your face if you follow Herbie's example.* **Lesson**: *How a thermocouple works and how it might not work.*

What would the 4th of July be without fireworks?

Herbie in the lab is almost a guarantee. Last week he brought a new module to the Environmental Lab for UL testing. It was the latest development from the Research Center in echo-canceling technology. It didn't seem good for the environment to just throw away all of those echoes, so they put together a new chip that strips the echoes out of phone calls, but instead of canceling them, it saves them, and sends them out to places that have a great need for second-hand, distorted information, like the Internet.

One of the tests UL likes you to do is to measure the temperature of certain components, such as transformers, while you vary the input voltage or short-circuit the output leads. For this test, Herbie borrowed my favorite data logger and a bundle of thermocouples. He glued thermocouples to all of the components that the UL engineer pointed at, hooked the other ends up to the data logger, and turned on the power. That's when I got the first phone call.

Hot Air Rises and Heat Sinks

"Your data logger is going haywire!" he yelled. I thought I could hear a smoke alarm going off in the background, so I hurried to the lab.

Everything was turned off when I got there. The UL guy was frowning at me, worried that the test would not be done by lunch time. "Show me what's wrong," I said.

Herbie turned on the data logger. It sprang to life, purring softly, its LED display glowing with a reasonable temperature reading. I patted its imitation wood grain chassis reassuringly.

"Hmmm. Seems OK now, " Herbie said. "You must have done something to it."

Then he turned on the power to the Echo Recycling Module (ERM). Right away the data logger started beeping in Morse code, the temperature display started flashing "666," and the printer began spitting out paper tape covered with the symbols they use in "Beetle Bailey" to denote swear words. I pushed the main breaker on the lab bench, and everything died.

My data logger was OK. It wasn't going haywire. The fact that it was going crazy in this case probably saved Herbie from collecting some very bogus temperature data. To explain this mystery, we need a little lesson in the theory of thermocouples.

Once upon a time, two guys discovered an obscure physical effect. Seebeck noticed that if you make a circuit like the one in Figure 16-1 out of two different kinds of metal (in his day, probably iron and pewter), and you heat up one of the junctions and cool down the other, you get a very small voltage (in the range of a few millivolts). Peltier got credit for figuring out that the circuit also works backwards — if you apply voltage to a circuit made out of two different metals, one junction gets hot and one gets cold. A third guy, Lord Kelvin, noticed that the voltage in the circuit was directly related to the temperature difference between the two junctions.

Now you've got all of the ingredients to make a pretty neat thermometer. You stick a piece of masking tape over the scale on your voltmeter and draw in a new scale of "degrees C" instead of "millivolts." It's not really that simple, especially if you want accuracy, but the principle is simple: to measure temperature, you really just measure voltage.

Thermocouples

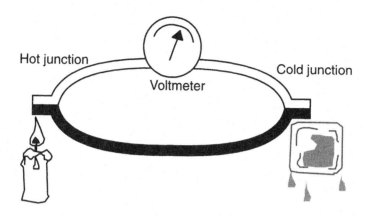

Figure 16-1 Two metals plus hot and cold equals electricity

That's what Herbie was doing. My data logger is essentially a fancy voltmeter — but a very sensitive one. He had glued one or two of his thermocouple junctions (which are usually bare metal to make a good thermal contact) directly to the windings of a transformer in a ring voltage circuit. Apparently the insulation on these windings is not so perfect (just a coat of varnish), and 90 or 100 volts were jumping onto the thermocouple wires and shooting into my data logger. The detection circuit that was looking for 20 or 30 millivolts and got smacked with 100 volts wasn't happy, which led to all of the sparks and alarm messages.

By the process of elimination we figured out which wires were shorted to high voltage and fixed the problem by insulating the thermocouples from the windings with a piece of polyimide tape about 0.001 inch thick. Kapton tape, a registered trademark of DuPont, giving about 7000 volts of isolation in that thickness, which introduced a small error in the temperature measurement, but allowed the test to continue.

Lessons that You Should Have Learned

You think you are measuring temperature, but you are really measuring voltage, and when you put your wire into a source of voltage, you might get more volts than you think. As previously hinted

Hot Air Rises and Heat Sinks

at, getting 100 volts in your data logger is not the worst thing that can happen. At least when the display is spinning like the digits on the National Debt Meter, you have a feeling the data aren't trustworthy. It is possible to induce voltage into your thermocouple circuit that won't cause your meter to jump, but it could give you a wrong temperature. For example, if you stick your wire onto the live tab of a TO-220 package that is at 5 volts, you might get a little current flow through the thermocouple wires back to the meter. Because the wires are made of different metals, they don't have the same resistance, and if you have current flow through two different resistors, there will be a voltage difference between them. A small voltage between the two wires of thermocouple is just what the meter is trying to measure, and it interprets this voltage as a false temperature reading. Your component may be at 30°C, and the meter could read 100°C. Or worse, the component may be at 100°C, and the reading could be 30°C. There are ways to avoid these voltage/temperature errors:

- Don't mount thermocouples directly on electrically live surfaces (these are not always obvious!).
- Use a meter that has good electrical isolation between the thermocouple measuring section and the rest of the meter. This prevents you from accidentally grounding the circuit you want to measure through the thermocouple wires.
- Check for weird voltage effects by turning off the power to the electronics you are testing and looking for large or unexpected sudden changes in the temperature readings. (Of course temperatures should go down when you shut off the power, but not by 100°C in 5 seconds.)
- Don't try to finish a temperature test just before lunch or a holiday weekend.

The last tip is the most important. It is the reason I said earlier that these are the lessons that *should* have been learned. Because it was after I solved the voltage problem for Herbie, and went back to my office to work on the next revision of the *G-D Guidelines*, that the real fireworks began.

Herbie was done measuring temperature and thought he could squeeze in one more test before lunch. This was the dielectric test, where you apply 1400 volts to various parts of the product to see whether the insulation is sufficient to keep it from breaking through to ground (like the product chassis). This is an important test, because it keeps you and me from getting electrocuted by lamps, toasters, and vacuum cleaners. The only problem was a little bit of haste. Herbie hooked up the 1400 volts, but forgot to disconnect the thermocouple wires, or even to turn off the data logger. The high voltage indeed found a path to ground, through the thermocouples, through the data logger power supply, and through its AC line cord.

At first he didn't even know what had happened, because there was no spectacular explosion, only a soft beep from the Dielectric Tester indicating an insulation system breakdown. Then he smelled the smoke coming out of the data logger and realized something was wrong. Then I got the second phone call.

That data logger has never been the same since. It wasn't completely destroyed, but now it kind of stutters and has little nervous tics. I suppose the only way I'll deal with my grief is to write about it, and to keep muttering to myself, "Never again, never again ... will I lend Herbie any of my equipment."

BUT THE GRAPHICS ARE PRETTY

CHAPTER 17

> *Computer simulation can predict the temperature of electronics even before they are built, and can be wrong to eight decimal places.* ***Lesson****: More on CFD.*

"Garbage in, garbage out." — *Oscar the Grouch*

After a few successes with The ThermaNator, the thermal simulation business began growing as fast as a sumac tree in a compost heap. Every new project plan now included a box labeled "ThermSim." I had to send out a memo to all of the project managers with the following warning: my interval for completing a thermal simulation is 1 week, but the schedule needs to include the couple of weeks the circuit designer takes to gather up all of my input data. My memo included the following tale to illustrate my point.

Herbie was in a bad mood and wanted to share it. "My project manager says we have to do thermal simulation on all of our new boards. I don't mind keeping you in business, but I don't see why I should bother, when you did such a — how shall we say? — less than perfect job on the PSI Module?"

"Oh, yes," I said, "the Psychic System Interchange Module for the HBU." I had run many variations on the layout of that module through The ThermaNator simulation program. Three high-power

Hot Air Rises and Heat Sinks

custom VLSI chips (code-named Id, Ego, and Superego) had to be tightly integrated or the system would go insane. This forced them to be very close together, and I had to juggle the layout and play with heat sinks until I found a combination that looked like it would work. Then Herbie built the PSI prototype and I measured the component temperatures.

I said, "As I recall, those Ego chips were within 5 degrees of what ThermaNator predicted. I'm considering writing a paper on that project for the ThermaNator International Users' Conference."

Herbie crossed his arms. "Don't forget to tell them about the AC11244 Line Driver. It didn't raise a blip in your simulation, but on the board it was so hot that it nearly burned a hole through your infrared video camera!"

"I can see why you weren't too happy with that, since you had to do another spin of the artwork to fix it."

Herbie grinned sourly as if he had just won a door prize that turned out to be an 8-track tape deck. "So what's the point in doing all this simulation? I could just build a prototype and test it, and then fix all of the thermal problems with another board spin anyway."

"I could explain this three ways," I said. "One is to blame myself, which is partly true, or to put the whole blame on you, which I probably could do, but it wouldn't do any good, or I could just say what happened, and maybe we can figure out how to do it better next time. Which do you prefer?"

He didn't say anything, so I outlined it the third way: The first thing you do in a thermal simulation is gather up all the information about the board you want to analyze: the Bill of Material (which is the list of all of the components and how many of each), mechanical drawings, component specs. The second thing you do is throw most of that information away. The trick is to not throw away anything important.

You throw away information because simulation is a *simplification* of reality. A circuit board might have more than 1000 components. There is no way to include all of them in a thermal simulation — and there really is no need (Figure 17-1). Most of those 1000 components are tiny chip capacitors and resistors that put out only about as much heat as a flea walking on a dog. They don't get hot,

But the Graphics Are Pretty

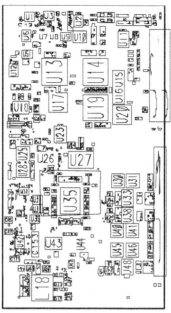

Original CAD layout of PSI board

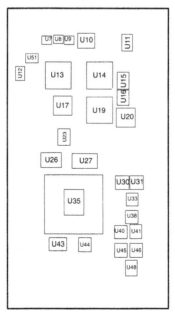

Simplified PSI model in ThermaNator

Figure 17-1 CFD can't handle all the detail of a real circuit board

Hot Air Rises and Heat Sinks

and they don't heat up other parts. Most of the power is given off by 10 or 20 components. These are the parts you need to know the temperatures of, the ones that cause all of the thermal trouble.

Step 2A is to pick out all of the high-power components from the Bill of Material. Step 2B is to look for temperature-sensitive components, like crystal oscillators. These components may not generate much heat, but they may have especially low operating temperature limits, so you want to know how hot they get anyway. Step 2C is to toss out everything else and start simulating.

This is where Herbie's AC11244 got lost. I saw it on the Bill of Materials, my red pen ready to strike it out, but then I took an extra step. I did not know the power dissipation of this part. My background and training is mechanical engineering — steam generators, dirty fingernails, and steel-toed shoes. To me, every electronic component looks like a resistor. Electricity goes in and heat comes out. A schematic is as indecipherable as an airline ticket price schedule. After a lot of rummaging around, I found that this same component had been used in another simulation, and 0.1 watts seemed to work well for it then, so I used it again.

The 0.1 watts did not seem like much compared with the 1.7-watt Ego chips on the other end of the board, so I decided to ignore it. I gave a list of other components I wasn't sure about to Herbie, and he supplied the power numbers for them. I plugged them into ThermaNator. A few hours later it spit out a beautiful color temperature map. The simulated temperatures looked good. Based on these results, Herbie ordered artwork and built prototypes. When we compared the pretty color temperature map to the image from the infrared video camera, there was one obvious difference. In the upper left-hand corner was a red, glowing blob that should not have been there, and its name was AC11244.

Herbie said, "If you had asked me, I could have told you that that part was putting out 0.9 watts, not 0.1 watts!"

I retorted, not very usefully, "If you knew it was 0.9 watts, why didn't you tell me in the first place?"

His mouth opened and closed a couple of times until an honest answer pushed its way to the front of his brain. "Because I didn't know it was 0.9 watts. I could have figured it out if you had asked me to, because it isn't that hard. You need the clock frequency, how many gates are used, and ... Anyway, we use this part all the time, and in a typical low frequency application, it doesn't get hot. At 10 MHz, it does run about 0.1 watts. But at 66 MHz, the power goes up 6.6 times to about 0.9 watts. I just didn't think about it."

"What we have here is a failure to communicate. I didn't ask and you didn't tell. One thing I'm starting to realize is that thermal simulation is not a push-button operation, no matter what those ThermaNator brochures say. There is no library or database where I just look up a typical power for any component. It depends on how the part works in the circuit. Somebody has to sit down and figure it out. Since you didn't, I made a guess, and I guessed low. What this shows is that maybe that's not such a good way to do it."

Herbie brightened. "So it IS your fault!"

"Let's just say 'human error,' OK? The point is that I have to put in the correct power if you want ThermaNator to have a chance at giving you a good temperature prediction. Remember this equation?"

$$Q = h A (T_1 - T_2) \qquad (17\text{-}1)$$

Or put a slightly different way:

$$\Delta \text{Temperature} = \text{Power} / (h \times \text{Area}) \qquad (17\text{-}2)$$

"Looks vaguely familiar," he said, "but I always get confused by that triangle thing."

"I know. That's a delta, which starts with a "D" and stands for "difference." Delta T means the temperature difference between the component and the air flowing by. This equation is one of the Ten Commandments of Heat Transfer. Even if you don't use it to calculate anything, it can tell you what things affect temperature. There are three terms: **Power**, **h**, and **Area**. ThermaNator figures out **h** (the convective heat transfer coefficient) for you, but the other two things you need to specify. **Area** is the surface area of the components and board, which I can get from a drawing of the board.

Hot Air Rises and Heat Sinks

Power is the heat generated by each component. The accuracy of the temperature prediction is directly related to the accuracy of the power you put in. If you have a 10% error in power, expect a 10% error in the temperature rise. If you have a 900% error, like with the AC11244, then you get a 900% wrong temperature rise prediction. Or in our case, I rounded the power down to zero, because it was so low I thought it could be ignored."

"But heat is *your* job," Herbie whined.

"Right. But I want you to know enough about it to understand why I need you to do some extra work before I can do mine."

"More work? I'm too busy already."

"Here's the new process for thermal simulation, which I am making up on the spot as of now: Questions should be answered by the person who most likely knows the right answer. I'll answer the ones about airflow through a shelf. But since component power depends on how it is used in a circuit, when we want to know the power dissipation of a component, the *circuit designer* — that's YOU — will answer that one. What do you think?"

Herbie said, "I like the old way better. I give you whatever data I feel like, and you still have to crank out the right answer in a week."

"We can work it that way, too. But here's the new deal. If I have to guess the power data myself, I'm not going to give you a pretty color printout of the temperatures. Only a table, and maybe a black-and-white temperature map."

"Hey!" Herbie protested, "you know I can't impress anybody without that color map! That's blackmail!"

"No, it's black-and-white-mail. Have we got a deal?"

We tried out the new arrangement during the simulation of the double-density PSI Module. The main problem was Herbie couldn't fit two Egos on one board. But he actually calculated the component power, keeping my guesswork to a minimum. With that process improvement, at least when the temperatures were off, we knew who to blame right away.

Too Much of a Good Thing

CHAPTER 18

Pin fin heat sinks look like they have a lot more surface area in those magazine pictures. How come they don't work any better? **Lesson**: *Convection works with surface area parallel to fluid flow.*

My wife and I were shopping at Hi-Tech Round-Up, the giant store near the tollway exit that sells music CDs and CD-ROM software with equal understanding. We were looking for a new modem for the home PC so we could send e-mail to all of the people we used to call from that phone in the airplane headrest and say, "Guess where we're calling from!" It didn't seem fun anymore since all those people got answering machines. In the computer doohickey aisle my wife pulled a blister pack off the rack.

"Isn't this the kind of thing you use at work? What is it you call it?" she asked.

I was a little shocked to see such a thing at a retail store. I suppose next they'll have lead-foil packs of plutonium.

I said, "You're right. It's a heat sink. But what I usually call that is a crime against nature."

Six or seven shoppers perked up their ears at that phrase. My wife rolled her eyes. "Is this going to be one of your speeches like Oliver Douglas on *Green Acres*? Why don't you write a book instead?"

Hot Air Rises and Heat Sinks

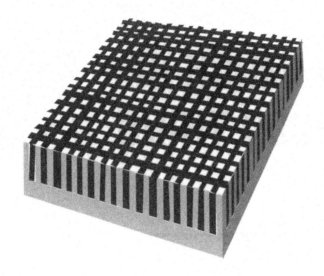

Figure 18-1 Pin fin heat sink: The choice of ignorance

So remember, it's *her* fault.

This is the crime against nature, the waster of aluminum — the ever-popular pin fin heat sink (Figure 18-1).

Hi-Tech Round-up sold them as "486 Coolers! Keep your PC's CPU from overheating PDQ!" On the bottom was a piece of double-stick tape. You just peel off the liner and slap it on top of your 486 microprocessor, and then you don't have to worry about premature failure of your computer due to heat.

I don't know how good a job it'll do inside a PC, but this is the pin fin Rule of Thumb I made up for TeleLeap:

Thumb Rule No. 1. DON'T USE PIN FIN HEAT SINKS.

I hate to give out Rules of Thumb. There's a reason for their name. The thumb is the clumsiest finger on your hand. I want designers and engineers to be subtle and clever and elegant and efficient in their thermal designs. But in the case of pin fin sinks, a simple rule is justified.

Based on a Good Idea

Heat sink vendors really push these things. They are featured in magazine ads and are on the front cover of the Sinkarama catalogs. The main reason pin fin heat sinks are so popular is that they look exotic. "Look at my board. It is very high tech. It has a *pin fin heat sink!*" Besides that, though, the pin fin sink is based on real science. Remember this old equation?

$$Q = h\,A\,\Delta T \qquad (18\text{-}1)$$

or

$$\text{Power} = h \times \text{Area}\,(\text{Temp}_{high} - \text{Temp}_{low}) \qquad (18\text{-}2)$$

In street talk, the more surface area a heat sink has, the better it works. This is another one of those Rules of Thumb for heat sink design, but as you'll see a little later, it doesn't always hold water.

The concept of the pin fin heat sink is to try to pack as much surface area into a given volume as possible. It certainly *looks* like there is a lot of surface area sticking out all over the place on the heat sink shown in Figure 18-1.

The other concept of this design is that it works in more than one orientation. Air is supposed to be able to flow through the fins no matter in what direction it flows (except perpendicular to the solid base, of course).

Let's compare the pin fin to the heat sink design it is intended to beat out — the extruded fin sink (Figure 18-2).

The extruded fin sink is made by squishing aluminum through a heat-sink-shaped hole and then chopping it to the right length, exactly the way the Play-Doh Fun Factory works. It can make fins in one direction only. The most common kind of pin fin sink is made by taking an extruded fin sink and crosscutting it with a saw or mill (Figure 18-3).

Now remember, the idea is to get more surface area than you had before. Take a look at the two sinks in Figure 18-3. Which one do you think has more surface area? It's not that hard to figure out for yourself that every time the saw makes a cut, it creates some new surfaces, but it also takes some away. If the saw cut is wider than the pin thickness, then there will be less surface area than before. If the

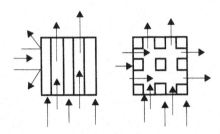

Figure 18-2 Extruded fins only work with airflow in one direction; pin fins work equally well in any direction

saw cut is narrower than the pin, then there will be a little more surface area. Commercial pin fin sinks seem always to have square pins, and the spaces are the same size as the pins themselves. With that kind of design, the amount of surface area is exactly the same as the original extruded sink.

Somebody should have said, "Hey, Vern, why are we spending money to crosscut this extrusion just so we get exactly the same amount of surface area?" But they didn't, and the story still isn't over. Because by itself, surface area isn't very useful. It has to be in contact with the air flowing by or it doesn't help at all.

The surfaces of the extruded fins are lined up with the direction of airflow (if you mount the heat sink correctly)(Figure 18-4). With the pin fin, only the vertical surfaces touch the flowing air. The horizontal surfaces touch only dead or recirculating air, and don't get rid of much heat. So not only is there no extra area, but now half of the area is facing in the wrong direction.

The pin fin inventors have achieved one of their design goals. Air flows through the pin fins equally in any direction. For natural convection (which is the mode for 90% of TeleLeap products) that means air flows EQUALLY POORLY in any direction. Without going into a five-page explanation of boundary layer theory, I'll describe the problem by saying that the pins on most of these pin fin sinks are just too darn close together. As air flows past a fin, it gets

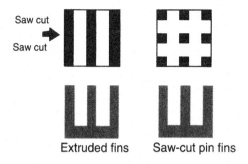

Figure 18-3 How to make an expensive, less effective heat sink out of a cheap one

disturbed. If two fins are too close together, these flow disturbances touch each other and plug up the flow channel with turbulence. The pins on the sink shown at the beginning of this chapter are so close together (about 1/16 inch) you might as well use a solid block of aluminum. The air just flows around the outside of the sink and over the top of the pins. The extruded sink from which it is made has the same problem. Its fins are too close together for natural convection flow.

Textbook equations can give you the optimum fin spacing based on the power and the length of the fins. For typical heat sink sizes (an inch or two wide by an inch or two long) and power in the range of a few watts, the right spacing is about 1/4 inch. Forget the equation. Remember 1/4 inch. Or better yet, if you can't fit the tip of your

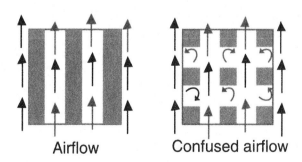

Figure 18-4 Pin fins put surface area where the flow is slow

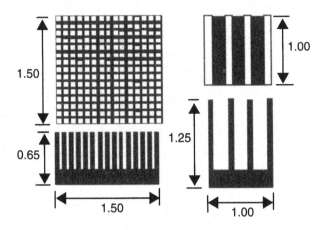

Figure 18-5 Which one looks high-tech, and which one works better?

pinky finger between the fins, don't use that heat sink in natural convection. Instead of a Rule of Thumb, call it the Pinky Rule.

A ¼-Inch Example

The Pinky Rule came in handy on the Optical Compensator Module (a sort of a set of reading glasses for a fiber optic receiver to help it detect signals that have come over quite a long distance and might be a little fuzzy). We needed a heat sink for a 3.5-watt integrated circuit in a package about 1 inch square. Figure 18-5 shows the two candidates.

Which one looks high-tech, and which looks like it would work better? Here are the actual case temperature measurements:

	Pin Fin	Extruded Fin
Surface area	30 in²	9
$\Delta T_{case\text{-}air}$	51°C	44

The extruded fin sink has less than one-third the surface area, yet it cooled the device 7°C better than the larger, more expensive pin fin sink. It just goes to show you that you can have too much of a good thing.

CHAPTER 19

COMPUTER SIMULATION SOFTWARE AS TEST EQUIPMENT?

Nobody trusts a computer simulation except the guy who did it, and everybody trusts experimental data except the guy who did it. Why not combine the two and get results everybody can mistrust a little? **Lesson**: *CFD as a way of interpreting temperature test results.*

The ThermaNator was pretty darned expensive. Every year when I wrote the purchase requisition to renew its license, I had to write an MOJ (Memorandum of Justification.) The standard way of thinking about ThermaNator was that you only use it at the very beginning of a new design project, before there is any hardware to test. You use it to estimate temperatures to decide on thermal feasibility. Every January, I had to face a series of managers with very reluctant pens for signing on dotted lines. On one side of the balance was my warm-and-fuzzy feeling of thermal confidence given by the ThermaNator, and on the other side was the $25,000 for the annual license renewal fee. If the balance needle swung the wrong way, I tossed on my hefty MOJ and got the money approved for another year.

ThermaNator As Test Equipment

Simulation was considered "nice to have," but not a "will die without it" item. On the other hand, nobody ever asked whether we

Hot Air Rises and Heat Sinks

needed thermocouples or an infrared camera. Temperature testing was a requirement. It was easy to justify money for test equipment, whereas simulation seemed so nebulous. So I wrote up a case study that showed that The ThermaNator should be put in the category of thermal test equipment. I argued that ThermaNator made thermal testing work better, and gave this example of real lab work to back it up.

The Case of the Homemade ZENO

Our VLSI department (those people in the dark room with thick glasses and very large computer monitors) came up with a custom chip code-named ZENO. I think this name came from its tendency to never quite reach the end of its development. They proposed putting it and its 2.8 watts into the itsy-bitsy package shown in Figure 19-1.

My job was to figure out whether a chip this powerful was even thermally feasible in such a small package in natural convection. It would need a heat sink, but just how big? And if it were really big, was there even a practical way of attaching it to such a small, delicate package? We had attached large heat sinks to other components with a big blob of thermal epoxy. That was OK when the heat sink and the component were roughly the same size, but I was worried

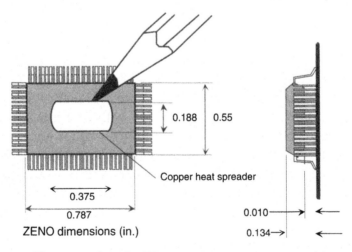

Figure 19-1 There was fear the ZENO package would not be able to go the distance

Computer Simulation Software as Test Equipment?

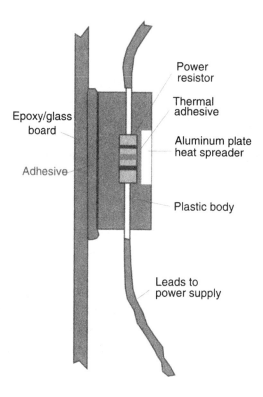

Figure 19-2 The homemade ZENO

that this sink would be nearly 10 times bigger than the ZENO, and that its large mass would yank hard on its tiny surface-mount leads. It was like tying up a houseboat to a dandelion stem.

There were no ZENOs available to play with at the beginning of the project, so I made my own. To represent the heat source of the chip I used a wire-wound power resistor. I bonded that to a piece of aluminum plate about the same size as the heat spreader in the ZENO package. I encapsulated that in the clear casting plastic that model railroaders use to make model rivers and lakes. To represent the circuit board, I used a — what else? — circuit board, except that it didn't have any copper in it. It looked like Figure 19-2, only messier.

This is what I had in mind. In the real ZENO, the 2.8 watts would split into two parts, each finding a separate path from the chip to the air. Some of it would conduct through the leads into the board, and

Hot Air Rises and Heat Sinks

some would go through the heat spreader, into the heat sink, and then into the air. The trouble is, I didn't know how much would go each way.

"So what?" I told myself. "You're not trying to duplicate exactly what the ZENO does. You want to test HEAT SINKS." So I purposely designed my homemade ZENO to force ALL of the heat toward the heat sink and none to the board. I thought I could do this by making all of the materials on the heat sink side good conductors of heat (like the aluminum plate), and by making all of the materials on the board side good thermal insulators (like plastic). Of course, some heat would leak through the plastic into the board, but the amount was bound to be so small that I could ignore it. Or so I thought.

My main goal in building this mock-up was to find a heat sink that could cool the real ZENO. I figured that if a heat sink would work on my homemade ZENO, it would work even better on the real one, because the real one could also conduct part of its heat away through its leads.

I started out to test a whole array of glues and tapes and heat sink shapes. At the same time, I simulated what I was testing with Therma-Nator. The first shape I tried, because it was easy to build and easy to simulate, was a flat plate heat sink, with the fake ZENO running at 2 watts. Here is how the ThermaNator results compared to the test data:

	ΔT Sink-Air	ΔT Spreader-Air
Test data	27.6°C	35.3°C
ThermaNator	33°C	44°C

For some reason, the simulation said that the heat sink and the ZENO should have been much hotter than I was measuring. With such simple geometry, ThermaNator should have been able to predict closer than this. So I started to look for sources of error (excuses) in my experiment.

In the MOJ, and in the paper I wrote for the ThermaNator Users' Conference, I quickly found the source of the error. In real life, though, I explored several dead ends.

Herbie noticed the odd fixture on my lab bench one morning while he was "borrowing" a nut driver from my toolbag. So I had to explain to him what it was for.

Computer Simulation Software as Test Equipment?

"Just how good of an insulator is that hobby shop plastic?" he asked.

"I don't know," I said. "All plastics have pretty much the same thermal conductivity. Probably around 0.2 watt/meter/°C."

"Is that the value you used in ThermaNator?"

I hesitated with my mouth open, suddenly realizing that it was never too late for a Thermal Guru to learn something. "Uh, no. I used zero. I made it a perfect insulator."

Herbie grabbed the wires that carried the current to the power resistor. "And what about these? These are copper wires. Shouldn't some heat leak out through them?"

"Well, a little," I admitted, "I ignored them, too."

I had assumed that the board side of the homemade ZENO would not conduct any heat. The plastic and glass/epoxy board were *good* insulators, but not perfect. The power wires to the resistor were small in diameter, but maybe I shouldn't have ignored them completely.

To look at these possible leakage paths I had been ignoring, I did one more test. I ran the homemade ZENO at 1 watt with no heat sink at all. This would eliminate as many unknowns from the problem as possible. Then I simulated the same thing with ThermaNator, this time putting in the actual values of thermal conductivity for the plastic and epoxy/glass, and the power leads. If the heat flow through the board was as small as I had assumed, then ThermaNator would confirm it. This table compares the thermocouple that was glued directly to the resistor heat source to ThermaNator's prediction of the resistor temperature.

	ΔT Resistor-Air
Test results	60.8°C
ThermaNator with no leakage	91°C
ThermaNator with leakage	56°C

ThermaNator told me that if I had really made my ZENO out of perfect insulators, as I had assumed, the resistor temperature should be a lot higher than I was measuring. My design assumed that 100% of the heat should flow through the heat spreader. In fact, in this test with no heat sink, more than 75% of the heat went out through

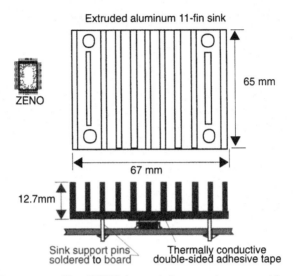

Figure 19-3 The ZENO heat sink turned out a tad large

leakage paths instead of the heat spreader — 75% versus 0% leakage was an important difference.

A Conclusion Worth Jumping To

ThermaNator helped me quantify the source of my experimental error. It told me I could ignore radiation and ventilation drafts safely. It gave me a way of testing my assumptions about how heat would flow in my homemade ZENO. I changed the design of my test fixture, which eventually led to the successful heat sink design for ZENO shown in Figure 19-3.

The sink did turn out very large, so large in fact that it needed four legs of its own for mechanical mounting to the printed circuit board.

This one real-life example was enough to convince me that ThermaNator was more than an ivory tower "what-if," nice to have temperature-predicting box. It had become a true piece of lab test equipment. It made test results more understandable, and "real" test equipment work better.

This MOJ proved to be either convincing enough or confusing enough that no one ever questioned my requisitions for ThermaNator money again.

THERMOTRIPLES

CHAPTER 20

Thermocouple folklore and the ongoing debate: Should you weld or solder your thermocouple junctions? It doesn't matter if you're going to measure in the wrong place anyway. **Lesson**: *More details of why thermocouples work.*

A manager in Herbie's department, a guy who really knows how to get the most out of TeleLeap resources, wrote:

```
Dear Uncle Thermal,
   I have run into a thermal problem
while fixing up an antique (1974)
snowmobile. The thermocouple that runs
from the air-cooled cylinder head to
the gage on the dashboard broke. The
original wires were crimped to a ring
terminal which fits over the spark
plug threads. The crimp barrel is
wrecked and I can't find another ring
terminal like this in the TeleLeap
stockroom. Is it OK to solder the
wires to the old ring terminal, or
will introducing a third material
```

screw up the reading? Please hurry. Winter is coming, and I just can't enjoy snowmobiling without being able to constantly check my cylinder head temperature.
— Hot Head, Milton Township

This was my reply, which I posted on the cafeteria bulletin board, along with the original note.

Dear Mr. Head,

You have touched on one of the most needlessly misunderstood topics in thermal engineering. Every thermal engineer I meet sooner or later asks, "Do you weld your thermocouples or solder them?" You have introduced a third method of making the junction between the two thermocouple wires — crimping.

Remember how thermocouples work: you make a circuit out of wires of two different metals. When there is a temperature difference between the two joints, or junctions in the circuit, a tiny voltage is generated.

The voltage in the circuit in Figure 20-1 depends on what the two wires are made of and the temperature difference between the two junctions. If you use calibrated wires, then the voltage is a way of measuring temperature.

This simple picture doesn't tell you what holds the two wires together at each junction. Real junctions need

Thermotriples

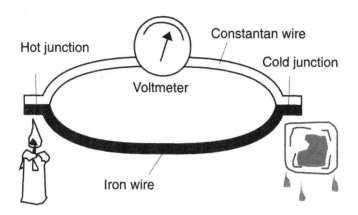

Figure 20-1 The ordinary, two-ended thermocouple

some physical connection: solder, glue, crimping, welding, or simply twisting the wires together. All of these methods have their own drawbacks, but they all have pretty much the same solution for reducing the error in measurement that they introduce.

When you solder or weld the wires together, what you are really making is a *thermotriple*, or a thermoelectric circuit with three different materials, which looks like Figure 20-2.

I have used iron wire and constantan wire just as an example. Constantan is an alloy of copper, which is commonly paired with iron in commercially available thermocouple wire.

What happens when you make a circuit like this? Because there are so many candles and ice cubes lying around my

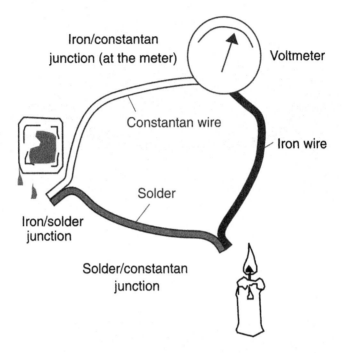

Figure 20-2 Three metals makes a thermotriple?

lab bench (sources of heat and cold), it is likely that the three junctions are at three different temperatures. You have an iron/solder thermocouple, an iron/constantan thermocouple, and a solder/constantan thermocouple all hooked up, and when the three junctions are at different temperatures, there will be three different voltages generated, possibly in opposite directions. This makes it difficult to know just what you are measuring. Another problem is that you don't know the calibration curve for iron/solder or solder/constantan thermocouples, and a typical temperature meter assumes the whole circuit is made up of only iron and constantan. Obviously, the temperature reading won't be right.

There is a way to get around this problem, and I did a little experiment in my lab just to prove to you that it will work. You can even do the same thing in your own lab if you don't trust me.

First, I made a circuit just like the one in Figure 20-3. Everything was at room temperature, including my coffee, which I had forgotten about during the test.

The thermotriple reads correctly when everything (both external junction and the one at the meter) is at room temperature. This doesn't really prove anything, except what we already know, which is that when all the junctions are at the same temperature, there is no voltage in the circuit.

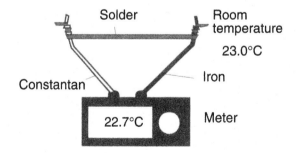

Figure 20-3 All three junctions at the same temperature

Hot Air Rises and Heat Sinks

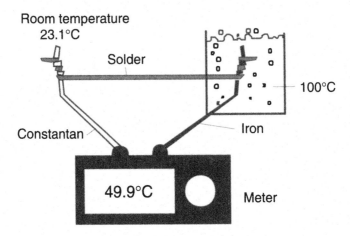

Figure 20-4 One junction at room temperature, the other at 100°C

Next, I put one of the junctions into boiling water, leaving the other two junctions at room temperature (Figure 20-4). I used water because it has a very well known boiling point, and boiling water is within even my skills. I thought maybe you wouldn't trust me if I just compared my thermo-triple against other thermocouples, as they have soldered junctions and might be inaccurate in your opinion.

Whoa! A reading of 49.9°C is a pretty big error, if what we wanted to measure was the water temperature. This trial shows what I said before from theory — once there are temperature differences among the three junctions, you start to get wacky readings. And they aren't mirror images. It matters which junction is hot and which is cold. Check out the next trial,

Thermotriples

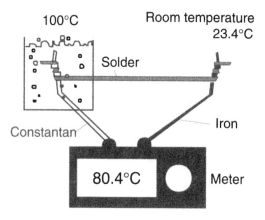

Figure 20-5 Reversing the temperature doesn't give the same reading!

where I switch the boiling water to the other junction (Figure 20-5).

This is still a pretty big error, and it doesn't fit with the explanation in the back of your mind that says, "Maybe it's just taking an average between the boiling water and the air temperature."

Now that we are convinced that solder in the circuit can give us really big temperature errors, let me show you how to get around it. Here is the last round of the test (Figure 20-6).

Is this what you expected? When I put both of the solder junctions into separate beakers of boiling water, I suddenly get the right reading, even though most of the solder itself is outside the water, pretty close to room temperature. How bizarre, and at the same time, completely predictable.

139

Hot Air Rises and Heat Sinks

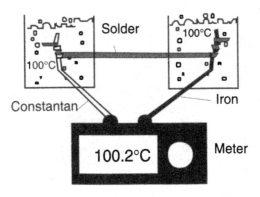

Figure 20-6 Make the two solder junctions the same temperature, and the measurement error goes away

When both solder junctions are at exactly the same temperature, they don't make any extra voltage, and the circuit works just as if the solder wasn't there.

But What About Real Life?

You have a choice of how to apply this new knowledge to your everyday thermal life. If you are going to solder or weld thermocouples, you can:

a. boil them to get the errors out; or

b. make your junctions small enough.

How small is small enough? I know you want me to say ¼ inch diameter, or 450 angstroms, or something definite so you don't have to think.

Thermotriples

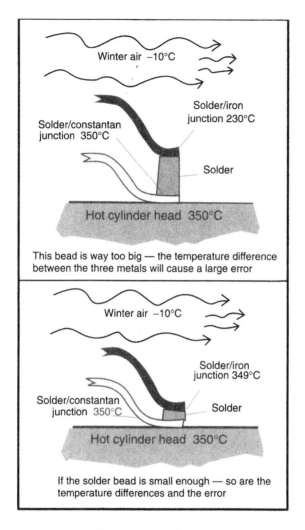

Figure 20-7 How this applies to snowmobiles and the rest of the real world

Figure 20-7 will give you an idea of the difference between small enough and too big.

The first thermocouple has a thick layer of solder that makes the bead so big that the top wire is sticking out into the cold air and is much colder than the wire touching the cylinder head. Because the two solder junctions are at different temperatures, they will cause big errors in the temperature reading.

The second thermocouple has only a thin layer of solder between the wires, and the bead itself is small enough that all of the solder is at practically the same temperature. Because both solder junctions are at the same temperature, the solder won't cause any errors worth worrying about.

Thus the final answer is that it doesn't matter whether you solder or weld or crimp, so long as you use a reasonably thin layer of solder, and that the resulting bead ends up being small enough so that the whole thing is at one temperature during your measurement.

I like to use 30-gage wire for my lab work and make beads about the size of the ball from a Bic medium point pen. This should be good enough for measuring most electronic components. For snowmobile engines, the beads could stand to be somewhat bigger.

Your idol,
The Thermal Answer Man

MIXED-UP CONVECTION

CHAPTER

Natural convection and forced convection should be friends. Why make them fight each other, unless fans of the Chicago Cubs are involved, and then nothing needs to make sense anyway. **Lesson**: *What happens when natural and forced convection work in opposite directions.*

"It's called VirtualFan," Herbie explained. The Sedona design group had finally come up with an application of their telepathic cross-connect technology that seemed to work well. "It's not very accurate with voice or data, but it responds amazingly well to the user's emotions, especially hope. Marketing sees it this way: sports fans are watching a big pay-per-view game on TV, and at the same time, they are hooked up to VirtualFan. It detects how much the fan wants his or her team to win and sends that signal to the central processor. After adding up the audience reaction, in real time, it signals the bio-pack on each player, which then releases either a mild tranquilizer or the hormone Hypertonin to cut back or boost the player's performance on the field. It is meant to recreate the effect that a live, cheering crowd has on the adrenaline level of the players. When we sell the pay-per-view, we also sell the idea, which most fans already believe anyway, that their rooting at home affects the outcome of the game."

Hot Air Rises and Heat Sinks

"I love it," I said with a complete lack of Hypertonin. "What I don't love is the mechanical design of the SportsFan System. The whole thing is thermally upside down."

Herbie shrugged his shoulders. "It's too late to change. The SportsFan box and power supply and a whole bunch of boards that do the pay-per-view interactive stuff are already finished. VirtualFan is just one new board that plugs into it. We can't change the fans or the vents or the size of the system box now, just because we're adding a new board."

"So if the design is all done, what do you want me to do?"

"The typical SportsFan board, which we have already built and tested and we know is OK thermally, runs about 5 watts. The VirtualFan board will put out about 20 to 30 watts. The HBU Interface chip is about 5 watts by itself, as it has to react faster than the mood swings of 64,000 sports fans. We need to run a ThermaNator simulation of this board to make sure we have a good layout for the components, and to help us pick out the right size heat sink for the HBU Interface chip."

"Thirty watts is a lot of heat for one board."

"Don't worry," Herbie said. "Remember, SportsFan is fan cooled. And there are two fans, in case one of them breaks. This should be a snap."

Herbie loaned me a SportsFan system box, complete with all of the boards that go into it, so that I could measure how fast the air flowed through it. It looks like Figure 21-1, from a thermal point of view.

The power supplies needed fans to stay cool. Somebody had decided, why not cool two birds with one stone, so they put the inlet air holes in the top of the power supply shelf, then stuck the system box on top of it. Air gets sucked into the open top of the system box, flows down through the SportsFan boards, out the open bottom of the system box, through the power supplies, and then out through the fans.

I measured the typical air speed in the channels between boards. It was about 100 ft/min, which sounds like a lot, until you realize that it is about 1 mile per hour. In ThermaNator, I made a computer

Mixed-Up Convection

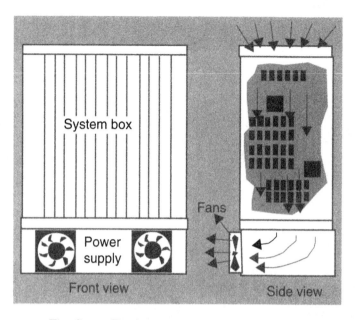

Figure 21-1 The SportsFan box had a very unnatural airflow direction

model of Herbie's initial layout of the VirtualFan board, plugged in the 100 ft/min air speed, and ThermaNator spit out its prediction of component temperatures.

The temperatures were nice and low, but that was with both fans running. The system was supposed to keep working even if one of the fans quits. I went back to the lab and disabled one of the fans and measured air speed again. This time it was only 30 ft/min.

An air speed of 30 ft/min was bad news. Why? Because 30 ft/min is about the same speed that air flows in natural convection. (In case you don't remember, natural convection is how a chimney works — the components heat the nearby air, and the hot air rises, starting an upward current.) Now you've got a SportsFan system with one dead fan, but you don't want to shut it off (you don't want the game to go off the air just because a bearing seizes up). The remaining fan is trying to pull air downward past the board at about 30 ft/min, but the heat generated by the board is trying to make the air flow upward at about 30 ft/min. Which is going to win?

Hot Air Rises and Heat Sinks

Textbooks call this situation *mixed convection*, because the flow is a mixture of natural and forced convection. I call it *mixed-up convection*, because whoever designed a downward flow direction must have been a little mixed-up about heat transfer. Normally when you have a fan moving the air, it is much more powerful than the natural convection currents, so you can ignore them. But when the two flows are about the same strength, and in opposite directions, you can't.

I plugged 30 ft/min into the ThermaNator model and it struggled for many hours to produce a solution. It is very hard to calculate mixed convection — it is like trying to predict whether a coin balanced on edge will fall over heads or tails. Figure 21-2 will give you an idea of what the flow looks like for the three kinds of flow: forced convection, natural convection, and the mixed-convection solution that ThermaNator eventually simulated.

In pure forced convection, the air flows all one way, in a fairly even distribution. In natural convection, the air also flows all one way (up), but the fastest air is usually near the hottest components. Far from the heat sources, the air may not move at all. In mixed convection, the air far away from the board flows generally downward, but the air near the board can turn around and start flowing up again, at least for a while. This can cause recirculating air pockets, in which the same blob of air flows past a component over and over. This is not a good way to carry away heat.

Herbie was skeptical of the ThermaNator results. The component temperatures in mixed convection were borderline. "Borderline" means the temperatures of some parts were very close to the operating temperature limit. It is a common and frustrating occurrence in the simulation game, because you don't have a definite pass/fail answer to motivate people to do something.

"How accurate do you think ThermaNator is in this situation?" he asked me.

This time I shrugged my shoulders. "I don't know. I never simulated mixed convection before, and the technical literature doesn't have any cases in which somebody designed a system where the fans flow against natural convection — I guess because we know that it

Mixed-Up Convection

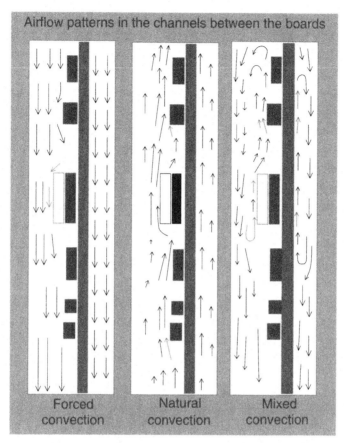

Figure 21-2 Two types of convection at the same time is not always an improvement

doesn't work as well as the other way around, and it is so easy to avoid. My recommendation is to mount the whole thing upside down and give the customers a big convex lens that inverts the image so they can read the silk-screens on the front panel."

I was looking forward to running thermal tests on the VirtualFan prototype. I wanted to see whether I could make the flow reversals visible with my smoke generator, and maybe write a technical paper about how good ThermaNator was at simulating mixed convection flow. But this was not to be.

Hot Air Rises and Heat Sinks

There was never any official story released about the field trial of the VirtualFan breadboard. Perhaps you heard the rumors. They hooked it up to 100 TV viewers of a Chicago Cubs game late in the season, and put the bio-packs on the players of both teams. None of the designers had caught on to the fact that the system had a positive feedback loop, which makes the whole thing inherently unstable. Although Cub fans are extremely loyal, nearly a century of mediocre performance has conditioned them into a pattern of low expectations. This game started off with three walks, which dulled the fan response, which made VirtualFan feed the Cubs tranquilizers, which made them slower, which bored the fans more, which doped-up the Cubs more, and so on. Soon nine uniformed guys were passed out on the field. After 20 minutes, the umpires finally decided something was wrong. Major league baseball, which had been desperate for anything that would make the game more exciting, called off the deal and later denied they had ever participated. VirtualFan died aborning, and took the SportsFan system with it.

It is just as well, I suppose. In more ways than one, if you want to get the most out of your fans, they ought to be in the right location.

A DEPENDABLE ANSWER

CHAPTER 22

> *How many watts can a 64-lead component safely handle? How big do the vents in my box have to be? What percentage of the heat comes out of the solder side of a printed circuit board? The answer to these and most other common electronics cooling questions is "It depends."* **Lessons:** *Component package power limits and their limitations.*

I spent a lot of time in the company cafeteria, but not for the food. It was just the best place to run into people and find out what they were working on. Because I was assigned to the Quality Simulation Department, a centralized corporate resource, I never heard about anything in the company except by accident.

I looked up from my bowl of chili to find my lunchroom table suddenly filled. Leading the pack was Herbie. "Speak of the devil," he said, "our Thermal Guy is right here."

Herbie introduced me to the two guys he was shepherding. Fergus was a custom ASIC designer from InteleLeap in Ireland, and Toivo was a circuit designer from Finland.

"We are working with Mr. Herb on a truly global project," Toivo said. "TeleLeap has licensed technology from New Zealand, the circuit design will be done in Finland and in the U.S., the boards will be built in Ireland, for use in a system engineered in the Netherlands,

Hot Air Rises and Heat Sinks

manufactured in Mexico, and sold in eastern Europe, the Middle East, and South America, to support automated closed-captioning of *The Flintstones* into 17 languages."

"Ah, global *and* noble," I said. "So how does a Thermal Guy fit into this scheme of Fred-and-Barney world domination?"

Fergus brushed the hair from his eyes and said, "I'm a bit new to TeleLeap and your way of doing things. Compared with my previous employer, MegaTechola, there don't seem to be many standards or rules for how to do a design."

"What kind of rules would you like to have?" I asked.

"Well, not to put too fine a point on it, but my job on this project is to design a set of custom chips, and I will have to decide how to partition the functions between them, how big the chips will be, how much power each one will dissipate, and what type of package the chips will go into. At my old company there was a written standard that said, for System XYZ, a PLCC with 84 leads will not have more than 2.7 watts, or something like that. I asked Herb where I could find the TeleLeap version of that standard. Much to my surprise, instead of giving me a document number, he mentioned 'this Thermal Guy'."

Toivo said, "Perhaps to you this sounds like a dumb question. But we would like to know what the thermal power limits are for different sizes of component packages. Are these numbers written down in a book somewhere, or do you have a Rule of Thumb?"

I winced at that expression and swallowed a mouthful of chili to cover the pain.

"I don't mind dumb questions, if you don't mind getting an equally dumb answer," I explained. "Not dumb, really, but maybe unsatisfying. How many watts can a 84-lead PLCC safely dissipate? The answer is that *it depends.*"

I let that hang in the air for a minute, noting their looks of dissatisfaction. "Let's look at it another way. Now you tell me — how fast can you safely drive your car on the freeway?"

Herbie shrugged, "I don't know, probably around 70 miles an hour." Fergus said about 60, and Toivo thought 80 was probably the maximum.

A Dependable Answer

"But what if it's raining?" I said, "Or snowing? What if the traffic is heavy, or even stopped? Is there any difference if you're driving a VW Beetle, a 16-wheeler, or a Porsche?"

That started a long discussion, with lots of stories about car accidents, speeding tickets, and how everyone else on the road is an idiot. "So it all depends, right?" I summed up, "How fast you can go without causing an accident depends on lots of local conditions."

Fergus said, "But surely there is a legal speed limit. That's the kind of rule we're looking for. A signpost to tell us when we're speeding, thermally speaking."

I said, "But the speed limit is somewhat arbitrary. It is based on what is safe for the lowest quality vehicle under ordinary conditions with an average driver. Ninety percent of drivers agree that you can usually drive at least 10% faster than the speed limit pretty safely. And some cars can go lots faster than that. On the other hand, just because you stick to the posted speed limit is no guarantee that you won't get in an accident."

"True enough," Toivo said with a nod.

Fergus said, "When you say, 'it depends,' what kind of things does the component power limit depend on?"

"Now that is a not-dumb question," I said. "Here's a list of things that affect the temperature of a component. Because that's what we're really talking about. You want to know how much power a component can dissipate before it gets too hot. And how hot a component gets depends on lots of factors." I wrote them down on a napkin.

- Natural convection or fan cooling: how fast is the air moving?
- Local air temperature: the hotter the air, the hotter the component
- Printed circuit board: how much copper it has and how big it is
- Neighboring components: how much power they give off, and how close they are
- Neighboring circuit boards: their power and how close they are
- Functional and reliability temperature limits of the chip
- Last, but not least: chip power and package size

Toivo squinted at the list, which spilled over onto a second napkin, and asked, "Do these factors have only a small influence, or are they large?"

Hot Air Rises and Heat Sinks

I said, "I can give you an example from one of Herbie's boards. This component dissipates about ½ watt and is used eight times on the board. The component at the top edge is 30°C hotter than the identical one at the bottom edge, just because the inlet air at the bottom is cooler than the air at the top. What kind of Rule of Thumb should I give you for that component?"

Toivo looked at Herbie unhappily. "This is complicated. You said this Thermal Guy would give us an answer. We need something simple to guide us in the design process."

I said, "I give you a complicated answer because I assume you want to optimize your design. Do you want your board to be bigger than it has to be, or more expensive, or only have half the features that it should have?"

"Of course not. It must meet all the customer requirements, and be minimum cost," Toivo said.

"To make a simple Rule of Thumb, I would have to be very conservative and give you very low power limits based on the worst combination of worst-case conditions. This would be like setting a 20 mile per hour speed limit all year because it might snow once in a while. To meet my conservative power limit, you might need to reduce your clock speeds, increase the size of your boards to spread out components, add fans and heat sinks, or maybe decide that the whole thing is just not feasible. But if we looked at your components in a complicated way, we might find that they will work without any of those changes. You want your project to be able to do 80 in a 65 mile per hour zone, don't you?"

Appealing to their manly pride was effective. "Darn tootin'!" Herbie said. "We don't want to follow any namby-pamby rules for baby designers."

"Very well, then," Toivo said, "but how do we proceed without design rules?"

"Instead of me giving you a single number to follow, you make me a part of your design team. I work with you through every step of your design, figuring out component temperatures based on all of those dependencies we already talked about. Here is where you start."

I whipped out the pocket-sized edition of the TeleLeap *Good Design Guidelines* (G-D) Manual. "On page 24, in Section 2.5.1.3.8, there are some tables called 'Package Power Limits.' That is your 20 mile per hour rule. If you are under the limits in those tables, you have nothing to worry about. But if you are over the limits, and if you are the heavy-duty designers I know you are you will be, but don't panic! Maybe you'll be perfectly OK at 35 miles per hour. At that point, get me on the case."

Natural Convection Power Limits

Tables 22-1 and 22-2 are based on the following conditions:
- Components are soldered into a printed wire board surrounded by similar components
- The board is mounted vertically and airflow is caused by natural convection
- The power limit is based on limiting the junction temperature to 70°C at an ambient temperature of 25°C
- Power dissipation has also been derated to take into account the effects of operation at 6000 feet altitude

Table 22-1 Through-Hole Packages

Package	Plastic	Ceramic
DIP 8	130 mW	130 mW
DIP 14	165	200
DIP 16	165	200
DIP 20	190	215
DIP 24	210	250
DIP 28	250	300
DIP 40	300	300
DIP 64	450	550
PGA 68		1000
PGA 84		1030
PGA 100		1100
PGA 149		1500

Table 22-2 Surface Mount Packages

Package	Plastic	Ceramic
SOIC 8	160 mW	190 mW
SOIC 14	180	200
SOIC 16	180	200
SOIC 20	225	270
SOIC 24	270	270
SOIC 28	250	300
SOIC 40	300	300
SOIC 64	450	550
PLCC 20	220	270
PLCC 28	240	290
LCC 44		320
PLCC 68	350	
PLCC 84	400	

Hot Air Rises and Heat Sinks

These limits will guarantee your components won't overheat. Follow them unless you need your product to be cheaper, faster, lighter, or smaller than your competition. Notice they don't even include many of the newer, popular package types. Does anyone still use through-hole parts?

Each of them feverishly wrote down the *G-D Guidelines* section number in their time management books. Toivo drew a picture of a speedometer, then wrote next to it 32 km/h.

Herbie jumped up suddenly and waved to someone across the cafeteria. "There's Spazz over there! I think we've got the thermal stuff we need for now. We have five minutes before our project management meeting. Let's get Spazz to give us the Electro-Magnetic Compatibility Rules of Thumb before then."

Common Questions

I have heard some thermal design questions over and over. They all have the same answer.

- How big do the air vents in my box have to be?
- What is the minimum space between circuit boards for good airflow?
- What percentage of heat comes from the component side of a board, and what percentage comes from the solder side?
- What is the smallest baffle that will work between shelves in an equipment rack?
- When the temperature limit from the *G-D Guidelines* is 125°C, what happens if my part goes to 126°C? 130°C?

Answer: *It depends.*

SUNSCREEN OR SMOKESCREEN?

CHAPTER 23

> *A university study claims that sunscreen keeps skin 20% cooler than bare skin. This is so obviously wrong that even an EE can spot it.* **Lesson**: *Temperature is not an absolute scale.*

The TeleLeap company nurse (the Medical Resources Department) hands out a free monthly magazine called *Health Hearsay*. It is filled with bizarre advice culled from various sources from around the globe. "Eat a carrot stick every time you make a cellular phone call. Not only is the carrot good for you," one item gushed, "but if you hang up as soon as you are done, it will help keep your phone bill down!" I was looking for an exciting new vegetarian recipe the other day, when I spotted a strange item. It announced that sunscreen did more than prevent sunburn. For bicycle riders, it can add moisture to the skin and helps cool it by convection. A university study showed that bike riders wearing sunscreen had a skin temperature *20% lower* than ones who didn't.

This was a real news story. Just remember: Free advice is worth every penny.

This sunscreen story sounds plausible on the surface. And quoting a university adds authority, too. It would take a thermal expert, such as someone who has read several chapters of this particular book, to spot

Hot Air Rises and Heat Sinks

the baloney here. Let's take a look at the thermal-related claims made in this tiny story and see how they are fundamentally impossible.

1. "Sunscreen adds moisture to the skin and helps cool it through convection." Convection has nothing to do with moisture. Convection is the process of air passing over a hot object and carrying heat away from it. There is no coating or gel or magic powder that will increase this process. It depends only on the speed of the air, the amount of skin exposed to the air, and the temperature difference between the skin and the air. I think they meant to write "moisture helps cool the skin by increasing *evaporation.*"

2. Evaporation is a totally different way of cooling. It takes a lot of heat out of your skin to change a liquid into a gas.

3. "Evaporation/convection, big deal, they got some terminology confused," you say. "They are still right that sunscreen keeps you cool." Wrong. Notice they said *moisture*, not water. That means that the liquid in sunscreen is probably some kind of oil or solvent, like alcohol. The solvent evaporates in a few seconds and is gone. Whatever cooling effect it had is gone before you finish putting the cap on the bottle. Oils don't evaporate very quickly at body temperature, and so they wouldn't provide much cooling.

4. The moisturizer in sunscreen works by permeating the surface of your skin with oils. This makes skin feel smooth and healthy. Plus, the layer of oil prevents water in your skin from evaporating. By preventing the evaporation of skin water (sweat), the moisturizer actually prevents you from getting cooler.

5. If you are riding a bicycle in the sun, you get plenty of convection from the air blowing by, and if that isn't enough, you'll start sweating and breathing hard. It is not unusual for an athlete to lose up to five pounds in a cross-country bicycle race. Most of that weight loss is water, given off as sweat from

Sunscreen or Smokescreen?

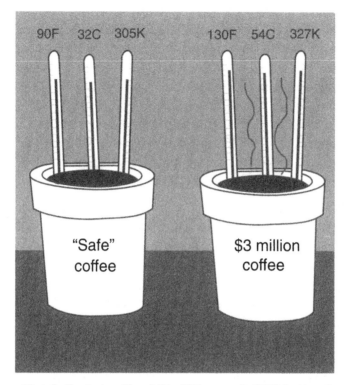

Figure 23-1 Is the hot coffee 44%, 69%, or only 7.2% hotter than the legally safe coffee?

the skin and vapor from the lungs. Compare a few pounds of sweat against the total amount of liquid in sunscreen. If you are generous, you maybe use a spoonful or two. Even if the moisture in sunscreen could evaporate (if your skin got to frying pan temperatures), it is nothing compared with the amount of liquid you would normally sweat anyway.

6. What about their experimental data? They claim that cyclists smeared with sunscreen have skin temperature 20% lower than cyclists that aren't. How many degrees is 20%? You and I have no way of knowing what that means. Temperature scales are relative, and it doesn't make any sense to talk about them in percentages or fractions or multiples (see Figure 23-1). By

Hot Air Rises and Heat Sinks

relative, I mean that they have an arbitrary value of 0 and 100. The centigrade scale is based on the freezing and boiling points of water, whereas the "F" scale had 100° as Gabby Fahrenheit's body temperature (maybe he had a fever that day). As for percentages or multiples of degrees, look at this example: if it's 10°C today and 20°C tomorrow, you might say it is twice as warm tomorrow. What if I convert that to the Fahrenheit scale? That's 50°F today and 68°F tomorrow. The same weather, and yet one thermometer says it is twice as hot, and another says it is only 36% hotter. To make it more confusing, what would tomorrow's temperature be if today's is 0°C?

7. Let's take a guess about what they mean by 20% cooler. Normal skin temperature is about 30°C, give or take about 5°C, which all depends on where you measure it and how hot the surroundings are. It is not nearly as constant as your inside temperature, such as under your tongue or in other places only medical doctors should be checking out. Let's assume that you are in the shade, at rest, and they measure the temperature of the skin on your arm. It might be 30°C. Then you hop on your bike across Death Valley in July. Your skin temperature goes up to 40°C. On the way back you wear sunscreen and it reads 38°C. That 2°C is a 20% decrease in your skin temperature change between resting and biking in the sun. Twenty percent seems like a big, unmistakable difference, but it is much smaller than the normal variation in skin temperature from moment to moment. That data is down in the noise. You wouldn't even feel the difference if it were true.

8. Here's another possibility: normal body temperature is assumed to be 98.6°F. After applying sunscreen and hitting cruising speed, your skin temperature goes down by 20%, to about 79°F. This would normally happen only after you jump into Lake Michigan. Call the paramedics. If you don't get help, you will die of hypothermia in a matter of hours.

9. This study compares the skin temperature of two different groups of people, not the same person with and without sunscreen. No two people have the same body temperature, especially measured on the skin. You know this can be a huge difference if your spouse's bare feet have ever bumped you in bed. From this news item you can't tell if they matched the screen-wearers and bare-skin cyclists with identical surface areas, clothing, age, body fat content, respiratory function, circulation, and political party.

10. How do you measure the skin temperature of cyclists while they are riding in a race? I suppose you could make them wear medical telemetry packs like the astronauts, but you would also have to measure the local air temperature, local air speed, incident solar radiation density, and relative humidity at all times so that you could factor out the other cooling and heating effects. Who would have hotter skin, the rider at the front of the pack, who is getting the maximum airflow, or the rider drafting behind him, who doesn't get as much cooling air, but who doesn't have to pedal as hard because there is less air resistance? I have a feeling this might have a bigger effect on skin temperature than sunscreen.

11. What does cycling have to do with this study anyway? (Could it be that the cycling contest was sponsored by a sunscreen manufacturer?)

When 70°C is Less Than 50°C

CHAPTER 24

> *Is a thermal test done at 70°C and 1000 ft/min air velocity more severe than a test at 50°C and 0 ft/min? Not always.*
> **Lesson**: *Convective heat transfer depends on a combination of air velocity and temperature difference, not just air temperature.*

"Robust is the new buzzword," Herbie explained. "New circuit boards have to be *robust*, like the flavor of mountain-grown coffee."

"A laudable goal," I said, "since the opposite of robust is *wimpy*. How do you achieve this robust-osity? Is there an IEEE standard?"

"Heck, no, it's much better than that. It's a new process that your department is pushing, called EST — Environmental Stress Testing. You take your new board to a special lab near Denver — "

"So that's where you were all last week," I said. "Your cube-mate said you were off skiing."

"Two birds, one stone — you know the theory. Anyway, this lab has a special chamber. You bolt the board inside and monitor its electronic functions while they subject it to all kinds of combinations of environmental stress. They run the temperature up and down, vibrate it like crazy, vary the input voltage."

"And then what?"

Hot Air Rises and Heat Sinks

Herbie continued, "They keep upping the stress levels until something breaks. They try to compress a whole lifetime of stresses into a couple of hours. It's a great way of finding out a product's weaknesses."

"Sounds great. Did EST tell you your board is robust?"

Herbie shrugged. "That's the confusing part. The EST manual tells you to stress the heck out of it until something fails, then you do what's called 'root cause analysis.' That means you try to figure out exactly what failed and what it would take to prevent that failure. Let's say the vibration makes a capacitor lead snap. You could fix that by changing to a different kind with shorter leads. But you don't always fix everything you find. For example, suppose you keep increasing the temperature, but the circuit keeps working and you get up over 120°C, and the epoxy/glass board starts to turn brown like a Thanksgiving turkey. We don't stop using epoxy/glass boards — we say that the stress was too severe."

"That makes sense. So what's your problem?"

"I'm trying to do the root cause analysis on my board, the SQUAT Module for the Gigantor System. The clock becomes unstable when they run the board in the chamber at 70°C or hotter. Vibration, low temperature, even temperature cycling are no problem. But go over 70°C and the clock gets flaky."

"Although I watch them all the time, clocks just aren't my specialty," I said.

"This isn't exactly a technical problem," Herbie said. "If I had the time, I could find the temperature-sensitive parts in the clock circuit and tweak them to get rid of the problem. But I don't have the time. And my boss says, hey, your board is only supposed to work up to 50°C anyway. It doesn't fail until it gets up to 70°C. That's 20 degrees of margin! Don't bother to fix it. There are plenty of other problems that need fixing more. Is he right?"

"Sounds like what you want to know is how 70°C air in the EST chamber compares to 50°C ambient in the Gigantor shelf. Right off the bat, I'll say those two things don't have much in common, so they'll be hard to compare. To start with, the Gigantor shelf is cooled by natural convection, so the air is just barely drifting past the components. The EST chamber blasts the air through the shelf at more than 1000 feet per minute. In the EST chamber, you are changing two variables on me at the same time: increasing the air temperature, which should make the parts hotter, but then also increasing the air speed, which tends to make the components cooler. Those two things fight each other. How the board works in the EST chamber at 70°C tells you nothing about how it works in the Gigantor shelf at 50°C."

Herbie said, "Yeah, OK, but which one is worse?"

"Both, is my best guess," I said. "Some components will be hotter in EST and some in Gigantor. It depends on the power dissipation of each component. Let's run a ThermaNator simulation on the computer so we can see. I happen to have the model we did of your SQUAT Module still on disk." With a few keystrokes and mouse clicks, I set up the boundary conditions to simulate the EST chamber.

Some hours later we were staring at these temperature maps side by side (Figure 24-1). Herbie was stunned. He said, "Is this realistic?

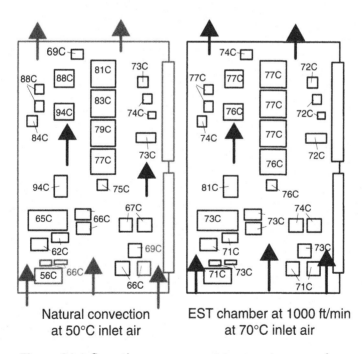

Figure 24-1 Sometimes component temperatures can be hotter at 50°C ambient than at 70°C

Hot Air Rises and Heat Sinks

Why do these two pictures look so different? How come the board at 50°C has components up to 94°C, and the board at 70°C only has components up to about 80°C?"

"The strong airflow in the EST chamber actually cools the hottest components and heats up the coolest components," I said.

Herbie kept staring at the pictures. "But which one is worse? I can't tell. The one on the left has the highest temperatures, but it also has the lowest. In the EST chamber, *all* of the parts are hotter than 70°C, but none get hotter than 85°C. Which board has the most temperature stress?"

"Neither one, or both, depending on which component you are talking about. Your boss thinks that because your board works up to 70°C in the EST chamber, you have 20 degrees of margin. But in reality, if the component that is causing the flaky clock is this real hot one up at the top edge of the board, it is running about 10°C hotter in the Gigantor shelf than it was in the EST test. Instead of 20°C of margin, you might be 10°C over the limit."

Herbie flinched at that idea.

"Of course," I said, "the exact opposite might be true. Let's say your flaky clock component is one of these low-power devices near the bottom edge of the board. In the Gigantor shelf, it runs about 53°C max. In the EST chamber, it goes up to 71°C. In that case, it is running much hotter than it would in operation, and you may have some margin. The only way you'll know is to finish the root cause analysis and figure out which component causes the flaky clock at high temperature. Then you can decide if it needs to be fixed. You can't just ignore the results of EST because the 70°C air temperature was higher than your operating limit."

"That EST guy told us that test would be useless without doing root cause analysis. I thought you could get me off the hook. You Quality Simulation guys all stick together," Herbie said, slinking away.

CHAPTER 25

EVEN A WATCHED POT BOILS EVENTUALLY

> *Roxanne the Intern hasn't learned her cooling folklore. Instead of following the traditional lab procedure of waiting an hour, then recording a temperature, she actually waits until the temperature reaches a maximum, and all heck breaks loose.* **Lesson**: *Thermal time constants and transient convection.*

To cultivate a respectable name in the academic arena, and to snag some cheap help, TeleLeap started hiring college kids to work as summer interns. They were a refreshing change from the veterans I dealt with because they still didn't know much, especially all the wrong stuff. One fine stifling day in early June, Roxanne appeared at the Thermal Lab, meekly returning a thermocouple meter that Herbie had borrowed last Christmas for two weeks. She stood in the doorway, staring at me with those wide eyes of innocent youth.

"What can I do for you?" I asked. "Did Herbie tell you to get the deposit back for the meter?"

"Maybe you can explain what I did wrong with this meter," Roxanne said.

"I like that attitude," I said. "Always assume you're wrong until you have no other choice."

Hot Air Rises and Heat Sinks

From her Iowa State backpack Roxanne pulled out a shiny new lab notebook. On page 3 was a neat perspective sketch of a transformer and several columns of figures. Then, she pulled out a dog-eared sheet of computer fan-fold paper, stained with coffee, a scribbled table of numbers running diagonally across it.

Roxanne explained, "Herb told me that this transformer is going to be an important part of the power supply he's working on for the Big Brother project. You know, that government deal so they can filter out all of the naughty bits from the Internet before it gets into the government computers. Anyway, he's worried that this transformer is going to get hot, and if it does, UL is going to make us spend more money on better insulation and all that, so he got three companies to make us samples. About a month ago, he measured the temperature rise on the first bunch. Then I got the job of measuring the rest."

"You don't have to tell me which results are yours and which are Herbie's," I said. I could almost see his image in the coffee stains, like some nightmarish Rorschach test. "Tell me what you did."

Roxanne continued, getting more nervous, "I glued the thermocouple wires to the transformer coils, just like Herb showed me. Then I hooked up the transformer pins to the power supply and the test load, the way he wrote in the procedure." She flipped over the fan-fold sheet and showed me a crudely drawn schematic of the test setup.

"OK."

"Just look at these temperature results, though!" she said, in distress. "My numbers are, like, way different from Herb's! I get coil temperatures around 90 to 95°C, and Herb only got about 70°C. I must have done something wrong, but I don't want to tell him until I know what went wrong exactly."

"Hmm," I hummed, and looked more closely at the two sets of data. Maybe the samples that Herbie measured were from a superior vendor that made a transformer that really did run a lot cooler. Roxanne had thought about that, too, and had already retested the same samples that Herbie had measured before — and came up with 95°C instead of 70°C.

Even a Watched Pot Boils Eventually

I questioned Roxanne about the details of her test setup: voltage, current, room temperature. She was sure she had followed the instructions Herbie had given her to the letter. And she had the detailed records to prove it.

"It's really hard to measure a *higher* temperature by making a mistake," I said. "Usually if you do something wrong with a thermocouple, like hook it up backwards, or glue it to the wrong spot, you wind up getting a colder reading. Plus, knowing Herbie the way that I do, I would suspect — Hey! What's this?"

At the top of each column of temperature readings was written the time of day. I compared Roxanne's and Herbie's data, then told Roxanne to relax.

We caught up with Herbie in his cube as he was sucking the last dregs of Nutrasweet from a can of diet cola. With a little encouragement from me, Roxanne showed Herbie the big differences in their measurements.

"Hokey smokes!" Herb said after he had digested the numbers. "Roxanne, how did you get these transformers so hot? According to this, my transformers will never work in the Big Brother system. And I already wrote in my status report that they would work great."

Roxanne's mouth opened, but nothing came out.

"Herbie, let me ask you a question." I butted in. "You want to know their steady state temperature, right?"

"Steady state?"

"You know, the long haul, full duration, till death do us part — the maximum temperature the transformer will reach after a very long time. When you did your test on the first batch, why did you stop measuring after only one hour?"

"I did?" Herbie said.

"Roxanne," I said, "why did you keep taking readings for almost six hours on each of your test samples?"

Roxanne cleared her throat and said, "Herb said I needed to find out the maximum temperature that the transformer coils would reach after turning on the power. I never did a test like this before, so I didn't know how long it would take to reach a maximum. Since we interns have a lot of time to kill, I just kept taking readings every

167

Hot Air Rises and Heat Sinks

15 minutes until it became real obvious that the temperature had stopped going up. That was usually between five and six hours. Sometimes I let it run overnight, see here on Trial 7, and then I'd look at it for an hour in the morning, just to make sure it was stable. I guess I didn't know any better, so I just sat around like a dummy waiting."

Herbie said, "Oh, yeah, now I remember why I used one hour as my time limit. You told me to. I needed to get some data fast for my status report, so I called you up to ask how long it should take to reach a maximum temperature."

"You did?"

"Actually, I left you voice mail, since you were off on vacation on that dinosaur dig. So I looked in my lab notebook from last year when we tested that HBU coil. It reached a maximum in less than 45 minutes. To get some margin, I rounded it off to an hour, took my readings, and wrote my status report."

"But that was a signal transformer about a tenth the size of this one! It didn't use a tenth of the power that this one does. Its thermal time constant is totally different," I said.

"Thermal time constant? You're just making that up. I think you heard that phrase on *Star Trek*."

"Not at all," I said, and took a piece of Herbie's fan-fold paper to sketch. "Let me explain. First, 99.9% of the time when we measure component temperature, we are trying to find out the *steady state* temperature, or how hot everything gets after a very long time, because our products usually get turned on once and stay powered up forever. It would be nice to know how long it takes to get to steady state so you don't have to sit and watch it forever. There are two ways to figure this out. Unfortunately, neither one of them is easy.

"If you know the power dissipation, the mass, and the thermal capacity of the component, then you can use this equation to estimate how fast the component temperature will go up. But the rate of increase depends on the temperature difference between the component and the air, so it isn't constant with time. The graph in Figure 25-1 shows what happens. When you first turn on the power, the temperature goes up pretty fast, but as it heats up, the rate of change slows down. The closer it gets to its final value, the slower it changes."

Even a Watched Pot Boils Eventually

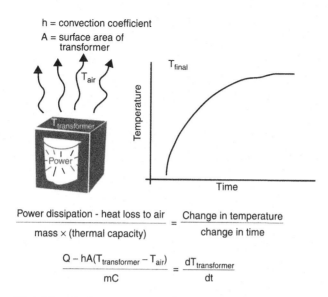

Figure 25-1 How fast a component heats up to its final temperature depends on its mass and thermal capacity, not how fast you need the results

Herbie frowned. "I don't like differential equations. Too many ds in them. What's the other method?"

"Roxanne's method. We pretend we don't know how to calculate the time constant of this component. We hook up your meter, turn on the switch, and just watch until the numbers stop changing. I'll give you a hint to get you started. In my experience, electronic components have a time constant somewhere between a few minutes and about 24 hours. And in this case, it's safer to use up some time and just keep measuring, than to be smart and stop before the temperature reaches its final value."

"OK, OK," Herbie said, "I goofed by trying to be too smart. It's a good thing Roxanne didn't know about my shortcut. Sometimes the ignorant approach is the best. Let me have your data. I'll need to update my status report for Big Brother."

We left Herbie grumbling in his cube. Roxanne said, "I still have more time to kill. Could you show me how to find the values of thermal capacity to plug into this differential equation? I'm curious to

Hot Air Rises and Heat Sinks

see how close it comes to predicting the temperatures I measured. I'm not afraid of equations with too many ds in them."

I loaned her a copy of my favorite undergraduate textbook, *Heat Transfer*, by J. P. Holman. The next day Roxanne returned with some computer printouts. "I solved this problem two different ways," she said. "First, I followed the analysis in the textbook and found a closed-form solution to the differential equation. I had to redefine some of the terms from the equation you gave us earlier:

$$\frac{Q - hA(T_{transformer} - T_{air})}{mC} = \frac{dT_{transformer}}{dt} \qquad (25\text{-}1)$$

"Since we already know **h** and **A**, then we can already figure out the final temperature of the transformer from the equation for convection heat transfer you keep showing us:

$$Q = hA(T_f - T_{air}) \qquad (25\text{-}2)$$

"Where I put **T_f** to mean the final temperature of the transformer. I can stuff this into the first equation and get that power term **Q** out of there. That makes the equation a little simpler. Or at least I could find one that looked like it in my differential equation book:

$$\frac{hA(T_f - T_{transformer})}{mC} = \frac{dT_{transformer}}{dt} \qquad (25\text{-}3)$$

"After you integrate this and find the constants of integration at the initial and final conditions, you get a pretty simple solution:

$$\frac{T_f - T_{transformer}}{T_f - T_{air}} = e^{\frac{-hA}{mC}t} \qquad (25\text{-}4)$$

"This equation looks familiar to us in the EE department, like the equation for how a capacitor charges up. That term in the exponent, **hA/mC**, is called the *time constant* of the equation. I guess that is the thermal time constant you were talking about yesterday. I knew there

Even a Watched Pot Boils Eventually

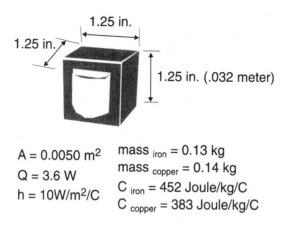

Figure 25-2 Roxanne's transformer details. (Sorry for the metric units, American readers, but they really do work much better in such problems.)

was one in the equation for an RC circuit. This one must be like the ratio of how much cooling there is to how much thermal mass there is.

"Here is what my transformer is like, along with how much copper and iron it is made of (Figure 25-2). It took me a while to figure out the amounts until I found out you could just call the vendor. They know, because every smidgen of it costs them something.

"From these values I figured out that **hA/mC** is about 0.00045 seconds^{-1}. The time constant is the inverse, about 2200 seconds, or about 37 minutes. In a circuit with exponential functions, they told us that in one time constant, the function reaches 63% of its final value. So without even doing any hard math, you can estimate how long you have to wait. After only half an hour, you would still be off of the final temperature by about 37%. After two time constants, you waited twice as long but you are still off by around 14%. After three time constants, you are up to 95% of the final value, and after four time constants, you have 98%. It's kind of amazing, but even without knowing the value of the final temperature, you can predict how long you have to wait before you will get within 2% of it. About four time constants, or 2½ hours."

Hot Air Rises and Heat Sinks

I was impressed. The only useful work I had been able to wrangle out of the intern in my department was getting him to make copies of all those back-issues of Dilbert for me. "Have you shown this to Herbie?" I asked. "It might be useful to him the next time he has to do a lot of transformer testing."

Roxanne flashed a proud smile. "That's why I had to solve it the other way," she said. "Herb seemed kind of confused by the differential equation. And he didn't like how the solution had that *e* in it. So I showed him how to do it on a spreadsheet with no calculus involved. I just took your original equation and rearranged it a little, and used discrete steps of time and temperature instead of the differential. Then you get a simple algebra equation like this:

$$T_{transformer,\ new} = T_{transformer,\ old} + \frac{\Delta t}{mC}[Q - hA(T_{transformer,\ old} - T_{air})] \quad (25\text{-}7)$$

"When you first turn on the power, the transformer temperature is the same as the air temperature, so the first value for $T_{transformer,\ old}$ is T_{air}. All of the rest of the terms in the equation are constants, so you can easily calculate the $T_{transformer,\ new}$. For the next time step, new becomes old, and you calculate the new temperature. Just keep repeating the same calculation over and over until the temperature stops going up. This is really easy to do in a spreadsheet. Here is a graph that Herb made himself" (Figure 25-3).

I was even more impressed. Roxanne had Herbie doing transient heat transfer calculations. Was there anything this intern couldn't do? "This is pretty good analysis," I said, trying to be encouraging. "Maybe you'd like to switch over to studying heat transfer in school."

Roxanne looked disappointed for the first time. "You know what lesson I've really learned on this project?"

"Tell me."

"All this stuff I've done is pretty much normal for what we do in college. We do labs and measure stuff and then use math and computers to figure out the answers to problems. And I like doing that. That's why I wanted to become an engineer. But it seems like once you get a job in it, you stop doing all that fun stuff. You spend all

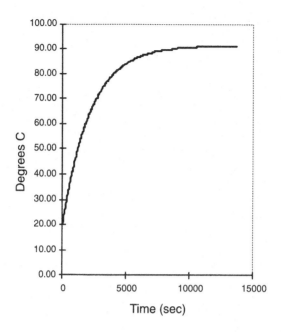

Figure 25-3 Transformer power-up

your time going to meetings and writing out forms and status reports. That's why Herb didn't measure the temperature right on these transformers. If he had time to think about it, he'd know how to figure out this time constant. Except he got so used to using short cuts, that he forgot the right way of solving problems. So what I'm really learning is, what is the point of studying all this engineering stuff in college if I'm never going to use it on the job?"

I nodded in agreement with her, thinking about the engineers I knew whose main job was to think up new acronyms to confuse their colleagues. Then I said, hoping to comfort myself as well as Roxanne, "You're right. Real engineering jobs aren't much like solving homework problems from the end of the chapter. You go to a lot of meetings, and you run into a lot of Herbies who have learned a lot of lame short-cuts. But once in a while, you do get to use your smarts to solve a good problem. That can be a lot more rewarding, at least inside your own gut, than those coffee mugs and T-shirts they hand out at the end of a long project."

"Well, OK," she said skeptically.

"And you know what else makes it worthwhile?"

"What?"

"Once a year, maybe, you find someone worth teaching a couple of things to."

THE LATEST HOT CD

CHAPTER 26

> *When you get a fever, the nurse doesn't have you put some ice under your tongue and then take your temperature again. Herbie wants to put a heat sink only on the components that have been measured as being too hot.* **Lesson:** *A complex assembly may have more than a single operating temperature limit, and the limit may change under different environmental conditions.*

In the good old days, a college dropout could write a program in less than 64 K (that's *kilo*bytes). That kind of economy is a joke now, when even a Garfield screensaver takes up 8 megabytes (MB) of disk space. I pondered this trend, wondering if I thought of computer memory the way my Depression-raised parents thought about cash. "Kids these days don't know the value of a byte!" So it didn't surprise me to find out that it took a 640 MB read/write optical disk to back up the system software for the HeadPhones administration unit.

Herbie showed me a sample of the optical disk and, as usual, the hardware amazed me. It was the size of a floppy disk drive, about 5 by 8 inches by half an inch thick. It used a removable cartridge that looks like a 3.5 inch floppy.

"And this thing can hold 640 megs?" I asked.

Hot Air Rises and Heat Sinks

Herbie grinned. "Not only that, but you can read and write to it as many times as you like. And it's fast — not like those tape cassettes that take half an hour to download."

"Wow. Looks like it fits into a personal computer. Can I get one for my PC?" I asked.

"Sure," Herbie said. "It's designed for the PC, but priced like something for the Pentagon. It costs more than your whole PC. But if you solve our little thermal question, maybe one of our prototypes will fall off the cart and get a scratch on the faceplate, so it would have to be, how shall we say —?"

"Misappropriated?"

"Whatever," he said, pulling a 150-page manual out of his bag. "Here's the thermal environment that OpticarTridge specifies. The drive is supposed to mount horizontally, needs a fan to blow 30 cubic feet of air across it every minute, and the inlet air is limited to less than 45°C. In our unit we mount it vertically, have no fan at all, and have inlet air up to 50°C. Will it be OK?"

I started laughing out loud. "You're serious, right? You want to mount a bicycle tire on a truck wheel, and I'm supposed to recommend proper inflation pressure?"

Herbie shrugged, "We know it's not rated to 50°C. But, as usual, we already told customers we would be coming out with this. Can't we just slap a heat sink on it someplace?"

"This isn't *Star Trek*," I reminded him, "I can't invent a new kind of particle that sends heat into a parallel universe. Leave the optical drive and the manual and I'll see what I can come up with. Sure you don't want to put a fan on this thing?"

"Sorry," Herbie said, "after the HBU incident with Vlad, we really want to stay away from fans."

The OpticarTridge manual was a rare object because it actually gave useful information, instead of just safety warnings. It was for people who build PCs. It not only told how much cooling air was needed, but it gave a method for telling whether you were cooling it enough.

The manual had a picture of the optical drive and its cartridge, marked with places to measure temperature, and a table of temperature limits (Figure 26-1; Table 26-1). Seemed simple enough. Put the

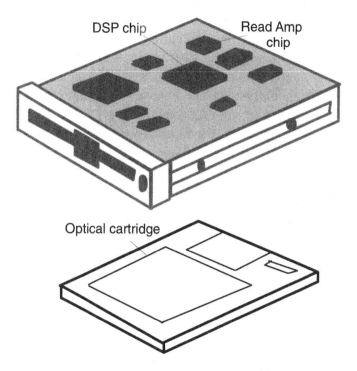

Figure 26-1 The vendor recommended measuring these spots to indicate the thermal health of the whole optical disk drive

drive in your PC, measure temperature in three places, and if you are below the limits, everything is cool. In more ways than one.

Table 26-1

Location	Temperature Limit
Read Amp chip	85°C
DSP chip	75°C
Optical cartridge	55°C

There was a glimmer of hope in this table. None of the limits was below 50°C, the maximum operating air temperature of HeadPhones.

I assumed that OpticarTridge had identified three appropriate places to measure temperature. If they had done it right, they would

Hot Air Rises and Heat Sinks

have taken the drive apart, put a thermocouple on every part inside, and then run it ragged. Then somehow they would have picked these three places to represent the temperature of everything. There were lots of components inside that optical drive, such as motors and lasers, that could get hot and behave badly when overheated. But the parts on the outside, such the Read Amp and DSP chips, should be more sensitive to how fast the air blows by than the stuff buried inside. After all, the wind has more effect on the bird sitting on my chimney than it does on me, all snug under the comforter in bed. But the picture and the table were all I had to work with.

So I stuck my temperature sensors on the three parts they recommended, put the drive in the HeadPhones Admin Box, and measured temperatures while reading and writing to the optical cartridge. Actually, Herbie did most of the operating and I did most of the fishing of thermocouple wires. Table 26-2 tells the story.

Table 26-2

Location	Measured Temperature	Temperature Limit
Inlet air temperature	50°C	-
Read Amp chip	81°C	85°C
DSP chip	73°C	75°C
Optical cartridge	65°C	55°C

"Looks like the weakest link is the optical cartridge. As the ambient goes up, it reaches its temperature limit before the other components," Herbie said, looking at the graph.

"Sorry," I said. "The cartridge is 10 degrees over its limit of 55°C, so it will never be able to work at 50°C ambient. Unfortunately, you can't add a heat sink or do anything to make the cartridge cooler."

Secretly I was relieved that the cartridge turned out to be the weak link instead of the Read Amp or DSP chips. If they had been too hot, Herbie would have asked me to recommend heat sinks to put on them. Of course, heat sinks could have cooled them below their limits. But then I would have had to explain why that wouldn't solve the problem. It would be like fixing the low oil pressure in your car by unscrewing the warning light bulb from your dashboard. The Read Amp and DSP chips weren't just hot parts — they were also indicators of other hot parts inside the drive you couldn't see. So cooling off the two indicators by putting heat sinks on them would only mask the problem. Hot parts inside would still have been too hot.

Herbie held a cartridge up to his eye and tried to look inside though one of the seams in the plastic. "Couldn't we drill some holes, or add a metal plate?"

"Herb, we don't even sell the cartridges to the customer. They buy them at OfficeJoint or CompMarket or Joe's Cheap Bytes. We can't make them run cooler."

"So what happens if they get over 55°C? They don't melt or anything, right? I ran one of these guys in the environmental chamber for three days at 60°C and it worked fine."

I said, "No, they don't melt at 55°C. From what I've read, they probably would work at temperatures higher than that. But the vendor stops guaranteeing them above 55°C. So they must see a certain percentage starting to act flaky, or not holding up over time, or starting to skip tracks, or who knows what kind of failure modes. Do you want to guarantee these cartridges above 55°C when the company that makes them won't?"

Herb nodded. "Lucky for us the customer won't be using this drive all the time anyway. Just for backups or loading new software. I'll talk to the marketing shamans and see what they say. Maybe we can put in the instructions not to use the optical drive when the ambient is over 40°C."

"Just don't make the print too fine, and I'll buy that," I said.

It turned out that the customers were willing to buy it, too. They were already used to the idea that the devices for backing up data don't work at 50°C. Before optical disks, they had to use tape cartridges, which happen to be made by the same vendors as optical disks. Above 45°C, the coating on the tapes starts to get mushy.

We did propose one mass storage device that would work at high temperature: a paper tape punch. It is nonvolatile, except around fire, and doesn't degrade over time, even up to 75°C. Unfortunately,

Hot Air Rises and Heat Sinks ─────────────────

640 MB of storage would require enough paper tape to give a parade to every astronaut that ever tried on a space helmet, and would take 18.2 years to load, not counting coffee breaks for the three shifts of machine operators.

Our customers settled for the optical drive, but marketing told us to keep working on the paper thing, just in case someone got picky.

WHAT IS A WATT?

CHAPTER 27

> *How hot does a component get that dissipates 1 watt? Like real estate, it depends on location, location, and location.*
> **Lesson**: *Convection + conduction = conjugate heat transfer, a tricky problem that can baffle your intuition.*

Joe from the Very Large 'Spensive Integrated (VLSI) circuits department dropped into my office. He is working on the Feature Anticipator (FA) chip. It was intended for those customers who can't decide what they want until after the product is completed.

"We thought we could do it with a simple flip-flop, but it turns out that the indecisiveness of our customers is much bigger than anybody thought," Joe told me.

"And I can help — how?" I asked.

"It looks like the FA chip will produce about 1 watt of heat. We aren't sure yet, but it will probably be in a plastic package that is about 1 inch square," Joe said. "I think about state machines, not steam engines. I don't have a clue how big a watt is. Is it going to be real hot or what?"

"OK, you want me to put it in scale, such as how many watts does it take to fill up the Super Bowl," I said. "First, let's look at a basic definition. A watt is a unit of power, and power is energy per unit time. And energy is the same as work. Here's an example. Stand up."

Hot Air Rises and Heat Sinks

When Joe stood up, I grabbed his baseball cap by the bill and tossed it from his head to the floor. Joe picked it up and as soon as he put it back on his head, I flipped it to the floor again.

"Hey, what gives!" Joe complained.

"A watt's what. If every 2 seconds you pick up your Kane County Cougars cap from the floor and put it on your head, you are doing work at the rate of about 1 watt of hat-lifting power."

Joe sat down and said, "A watt seems bigger than I thought."

I said, "I'm not counting all the wasted effort of you moving your body up and down, only the useful output of lifting the 4-ounce hat. Here's another number for comparison from Marks' *Standard Handbook for Mechanical Engineers*, 7th Edition. Just sitting there doing nothing, with your hat just sitting on your head not moving, you are giving off about 60 watts of heat."

"60 watts!" Joe said. "So how come I don't feel hot?"

"You would if the air wasn't constantly carrying the heat away in a huge, invisible plume," I said.

Joe shut his eyes and tried to work with the new ideas. "OK," he said. "So a human body is about like an average light bulb, although you'd swear from the look in some peoples' eyes, they must be down to 20 watts. And the FA chip is 1 watt, so it is about one-sixtieth of me. So how do you figure out how hot the chip will get — divide my body temperature by 60?"

I grit my teeth and said, "Not quite that simple. It depends on what is going on around the chip. There's a great illustration of this right across the hall from my office. When they cubified this building, they knocked out a lot of the internal walls and rearranged the air-conditioning vents. They managed to build this conference room with no ventilation ducts. Air can get in and out only through the door. I'll show you."

We walked across the hall and sat down in the empty conference room. "Seems OK to me," Joe said.

"Yeah," I said, "the two of us could sit in here all day with the door open. But try a two-hour meeting with a dozen people and the door open only a crack. You'll be sweating like Patrick Ewing at the foul line."

What Is a Watt?

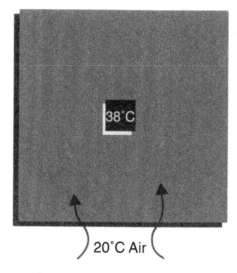

Figure 27-1 One watt all by itself in natural convection

Joe nodded and said, "So the temperature of my chip depends on its neighbors and the air flow?"

"Bingo!" I said. "Now let's look at some stuff on the computer."

I used The ThermaNator simulation program to make a model of a single circuit board with a single 1-watt component right in the middle. "No fans, right?" I asked. Joe nodded, so I simulated the board as if it were hanging by a string in the middle of a room full of 20°C air. Figure 27-1 shows the result.

Joe said, "My chip is only 18°C hotter than the air. That's not bad at all."

"Let's see what happens when we add a couple more of the same component, as far apart as possible. That way they shouldn't affect each other too much," I suggested (Figure 27-2).

Joe said, "The original chip hardly went up at all. Why are the new ones in the corners so high?"

I said, "Heat spreads out into the board from the component, as well as directly into the air. The board acts like a heat sink. In the corners of the board, heat can only spread in two directions instead of four directions."

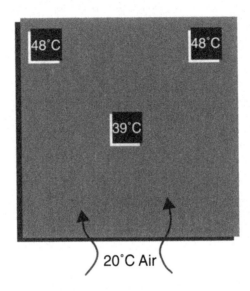

Figure 27-2 Each one of these is 1 watt, too

Joe started to get into the game. He said, "But what happens if we add two more, in the bottom corners? Would they be the same as the top corners, or would the air be cooler at the bottom?" (Figure 27-3.)

"It's hard to guess when you have all of these things that affect temperature combining in opposite directions. But The ThermaNator can do the 'What If' for us," I said. "Let's try it."

"This is weird," Joe said. "The lowest temperature is in the middle of the board. I would never have guessed that."

"It makes sense after you see the results," I said. "But it is hard to predict ahead of time. People just don't have good intuition about heat and temperature, even though we sense them all day long. What do you suppose would happen if we kept the number of components the same, but just pushed them all close together in the middle of the board? It is still only a 5-watt board, about the same power as a typical board in the Model 1 Crosser System." (Figure 27-4.)

Joe looked at the new results and said, "The center one got a lot hotter, as I expected, but the ones on top actually went down a little

What Is a Watt?

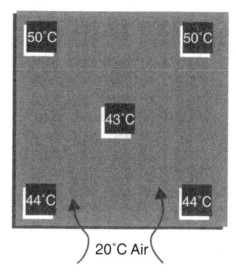

Figure 27-3 Why is the lowest temperature in the center of this 5-watt board?

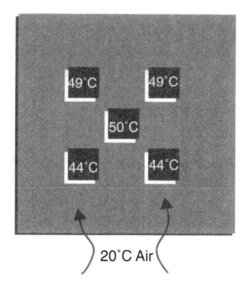

Figure 27-4 This 5-watt layout has the opposite pattern

Hot Air Rises and Heat Sinks

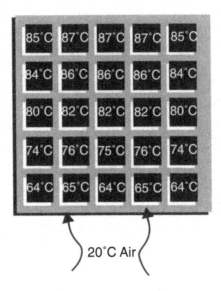

Figure 27-5 Is a 1-watt part 64°C or 87°C?

bit. Seems like you could play with this all day and get any temperature you want."

I said, "That *is* what I do all day long — draw board placements and see how the temperatures turn out. Let's try a worst case: fill up the whole board with 1-watt components." (Figure 27-5.)

"Holy Gallium Arsenide!" Joe exclaimed.

"Yep, these 87°C chips would probably be too hot to work for long," I said.

Joe looked like an orphaned pup. "So if a 1-watt part can be anything from 39°C to 87°C, how are you supposed to know what it really is?"

I answered, "The PC board designers don't call it *real estate* for nothing. To know how hot a part will be, you have to consider the three most important factors from the real estate game: location, location, and location."

RESISTANCE MYTHOLOGY

CHAPTER 28

Finding the junction temperature is the key to everything. But it turns out that the only way of calculating it is based on ancient mythology instead of physics. "But all legends have some basis in fact," as Captain Kirk says, so maybe you just stick with the myth until something better comes along.
Lessons: *Conduction; definition of the thermal resistance between junction and case.*

Word Problem: 75 people from your department must go from Tornadoville to Mallburg for an all-day seminar on "How to Plan More Efficient Meetings." Each person might drive a different kind of car. Some may carpool or use public transit. Each carload can choose its own favorite route, diverting to, say, Skinny Mama's Soyburgers for a bite to eat on the way. Your job as expense technician is to estimate the number of gallons of gas these 75 people will use in the nominal 15-mile drive.

Put away your maps and abaci. This problem does not have a solution, at least not a single correct one. Because you don't know how many cars will take any of several routes of different length and allowable speed, there is no way to calculate accurately the amount of fuel that will be turned into greenhouse gasses in the process.

Hot Air Rises and Heat Sinks

Yet, this is exactly the problem I had to solve every day. It is called "Figuring Out the Junction Temperature of a Component."

Herbie and others at TeleLeap used to judge the thermal righteousness of their designs by the exit air temperature. But I convinced them that it is not the air temperature, but the junction temperature (T_J) that determines whether the circuit will work. Perhaps they would not have believed me if they knew how tricky it might be to find the junction temperature.

This emphasis on junction temperature was so true and so useful, that even component makers like Motorola started to rate the performance of their components in terms of T_J. There was only one drawback to this belated step forward in logical thinking: it is nearly impossible to measure junction temperature. It is like measuring the temperature of your heart, except that cutting into your chest cavity and taping a probe to your beating heart would have far fewer problems than cutting into a component package and gluing a probe onto the silicon chip.

So my practice was to do the next best thing — glue a thermocouple to the outside surface of the component and measure *case temperature* (T_c). This is not hard, requiring only ten minutes of training in not gluing your fingers together. The junction temperature is found using this equation:

$$T_J = T_c + Q R_{j\text{-}c} \qquad (28\text{-}1)$$

where **Q** is the power dissipation of the component and $R_{j\text{-}c}$ is the thermal resistance between the junction and the case of the package. This number is supposedly a physical property of the package, depending only on its materials and construction details. Its value can sometimes be found in manufacturers' data books.

But $R_{j\text{-}c}$ is a myth. Like most myths, it has its origin in the beliefs of ancient peoples, in this case, component engineers of the 1960s. A clue comes from the very name "junction temperature." Why not chip or die temperature? Today there are dozens, hundreds, or even millions of junctions on a die. Junction is the term for the center of a transistor, the place where the transisting takes place

Figure 28-1 What a transistor looked like when $R_{j\text{-}c}$ was invented

and the Ns and Ps get together. The idea of junction-to-case thermal resistance was invented when a single transistor looked like the one in Figure 28-1.

This component was a hunk of silicon chipped off a boulder and crammed into a metal can. Heat was generated at the junction, which flowed through the silicon to the metal can and then into the air. The metal can was a pretty good conductor of heat, so it was all at one temperature. This allowed our engineering forebears to invent a simple method of calculating the junction temperature from the case (or can) temperature. They drew a thermal circuit, since they could only understand an equation if it looked like Ohm's Law, saying that power (**Q**) was the same as current (**I**), temperature (**T**) was the same as voltage (**V**), and thermal resistance (**R**) was the same as electrical resistance (**R**) (Figure 28-2).

There was only one path for heat to flow through, and so it was realistic to describe that path with a single resistance. Because this method was so successful, we got stuck with it. Even though components have become tremendously more complicated than the old tin-can transistor, we still pretend that there is only a single path for heat to escape. But Figure 28-3 shows what a typical plastic-packaged

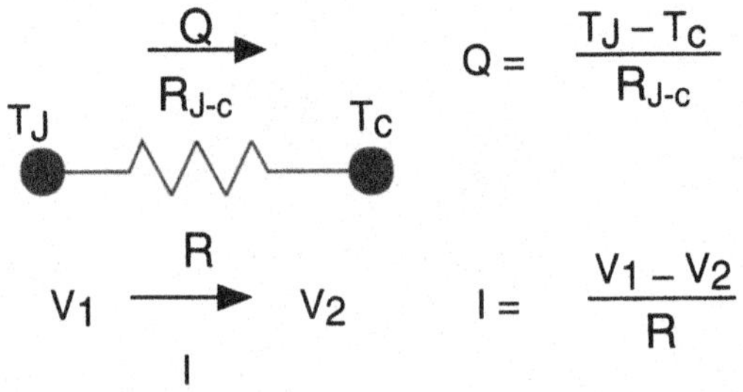

Hot Air Rises and Heat Sinks

$$Q \rightarrow \quad Q = \frac{T_J - T_C}{R_{J\text{-}C}}$$

$T_J \; \underset{\sim}{R_{J\text{-}c}} \; T_C$

$$V_1 \xrightarrow{R} V_2 \qquad I = \frac{V_1 - V_2}{R}$$
$$\;\;\;\;I$$

Figure 28-2 The heat conduction equation looks suspiciously like Ohm's Law

component looks like today. How could a single resistance correctly describe all of these different flow paths?

The little arrows represent how heat can flow out of the chip through many parallel paths. Not only does it flow through the plastic to the top of the package, but through the leads into the circuit board, out through the sides, and even through the bottom.

This problem was recognized right away, and over the years some well-meaning people came up with a brilliantly original concept for capturing this complexity, which is shown in Figure 28-4.

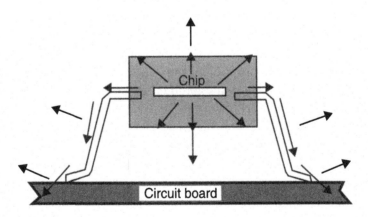

Figure 28-3 How could a single resistance account for all of the different heat paths inside a component package?

Resistance Mythology

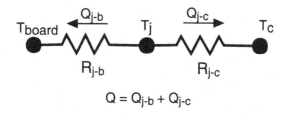

$$Q = Q_{j\text{-}b} + Q_{j\text{-}c}$$

Figure 28-4 A slight improvement on the $R_{j\text{-}c}$ concept

They took the beloved **$R_{j\text{-}c}$** model and grafted on a second path. They assumed that heat flows in two directions, from the "junction" to the top of the package (case), and through the leads into the circuit board. The resistance between the junction and the board was called **$R_{j\text{-}b}$**. They even published their values of **$R_{j\text{-}b}$** for some common types of packages. If you know the board temperature under the component, and the case temperature, you could calculate the junction temperature. This idea was a step in the right direction, but it never caught on, probably because of at least one tiny flaw:

The value of **$R_{j\text{-}b}$** is not really a constant, but depends heavily on how much copper is in the circuit board and how fast the air is flowing.

In the meantime, research has been going on to fix this problem. Clemens Lasance and Harvey Rosten (whom I tend to believe as I had dinner with them) wrote a paper on the subject.[1] They said that it takes more than one or two resistances to describe the heat flow inside a component. Figure 28-5 is an example of the resistance network they propose for a 208-lead PQFP (Plastic Quad Flat Pack).

Lasance and Rosten have even proposed a couple of experimental methods for measuring the values of these resistances so that they are independent of the environment and the construction of the circuit board. This approach looks very promising, but it is far from an industry standard yet.

---

[1]Lasance, C., Vinke, H., and Rosten, H., "Thermal Characterization of Electronic Devices with Boundary Condition Independent Compact Models." *IEEE Transactions on Components, Packaging, and Manufacturing Technology*, Part A, Vol. 18, No. 4, December 1995.

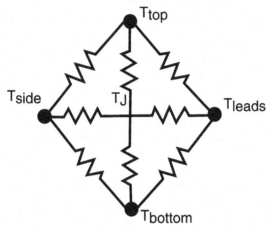

Figure 28-5 This is how many resistances it takes to get a practical model of a 208-lead PQFP

The reason I am showing you this spider web is to give you an idea of how complicated it *should* be to calculate a real value for the junction temperature. It is just as tricky as the gas mileage word problem at the beginning of this tale.

"But all legends have some basis in fact!"
— J. T. Kirk, USS Enterprise

If junction temperature is so important, and yet the best way we have for calculating it is nothing more than a myth, what are we supposed to do? I hate to say it, but as we don't yet have a practical and correct way of finding junction temperature, I am sticking with the Myth of $R_{j\text{-}c}$. I am not advocating that we bury our collective head in the sand and ignore the flaws of this method. I continue to use $R_{j\text{-}c}$ to find the junction temperature, recognizing that it is only an *estimate*. Until something better, such as the spider-web resistance method of Lasance and Rosten, comes along, I make use of $R_{j\text{-}c}$ because: The case temperature and junction temperature are related to each other *somehow*. I just can't be sure how much hotter the junction is than the case. So measuring case temperature is probably a lot better way of knowing the thermal health of a component than, say, the nearby air temperature.

- If I use the worst-case power to estimate the junction temperature, and if **$R_{j\text{-}c}$** is conservative, then the estimated junction temperature will also be conservative — that is, I will think the junction is hotter than it really is. This causes me to make the mistake on the side of safety (I think).
- For reliability I derate the operating temperature limits set by the component manufacturers. They say a typical die won't burn up until it hits 150°C, but the *Good Design Guidelines* says I can only run it up to 125°C. This derating allows me to be a bit off in my estimate of the junction temperature and still be safe.
- The boards I have tested by this method seem to be working well in the field. In the long run, that is what counts. (Although I wouldn't promise we actually have the right data to prove this.)

In the Middle Ages astronomers thought that the sun and stars revolved around the earth. Although the idea behind it was totally wrong, they had worked out a system that could still predict the positions of the planets. It wasn't right, but it worked, sort of. That's where we are today in the art of finding junction temperature — hit or myth.

THERMOELECTRIC COOLERS ARE HOT

Electrical Engineers love these all-electronic refrigerators. Herbie proposes them for use in a new system until he learns that not only do they cost a lot, but they still require fans and heat sinks, and leave the components hotter than they would be without them. Why are they so bad if they do everything the manufacturers claim? **Lesson**: *Peltier effect cooling.*

I heard there would be doughnuts at the All-Day, Let's-See-Where-We-Are Review of the MMMnMM (Multi-Media Mix 'n' Match Manager) Project, so I snuck in and sat in the back. I woke up when Herbie displayed a viewgraph of the system rack.

"We were able to squeeze an extra shelf into this rack," his amplified voice proudly announced, "by getting rid of the fan trays from the original concept."

I nearly upchucked my cranberry muffin in surprise, but I continued to listen. "We've decided to go with some new technology," he said, "getting away from old-fashioned mechanical heat sinks and fans, to something clean, reliable — and *electrical*. It's called the Thermoelectric Cooler."

Nearly everyone in the room turned toward me at the sound of the foam coffee cup crunching in my hand. The Managing Cheerleader (MC) of the MMMnMM Project leapt to the overhead

Hot Air Rises and Heat Sinks

projector, saying, "Didn't mean to step on anybody's toes, thermally speaking. But Herb told us about this amazing new device last week at the marketing meeting, and we just couldn't resist. It's a fully electronic device, no moving parts, that you can glue onto a hot component. It sucks the heat out and converts it back into electricity, which is then safely drained away to ground. Expensive — you bet! — but you never have to change an air filter or oil a bearing."

"OK," I said. "Thermoelectric coolers will not only cost a lot, but you'll still need fans and heat sinks, use more electrical power, and the components will be hotter than they would have been in the first place. Other than that, I'm all for them."

Herbie scowled. He suggested the meeting go on to the next agenda item, then he and I slunk out to an empty conference room down the hall.

"Here is the problem we were trying to solve a couple of weeks ago," I said. "You have an INFERNO chip like this on every board in the system. It gives off 4 watts of heat, and I calculated that it would get up to about 100°C unless we added heat sinks or fans." (Figure 29-1.)

Herbie said, "And there isn't room for heat sinks, because the boards are too close together. So you said we needed a big fan box in the rack. But we hate fans."

"Me, too," I said. "Except for almost every other kind of cooling device, I hate fans the most. I love thermoelectric coolers. They seem like a miracle. And they are. But it's like that Monkey's Paw story, where it grants your wishes but you wind up paying for them in ways much worse than you'd ever imagine."

"So the thermoelectric cooler is a scam, like that Crop Circles Futures Fund I put all of my money in last year?" Herbie said.

"No. I'm sure it does everything the sales brochure says it does. The technology is only a couple of hundred years old, so they've worked the bugs out of it. But I don't know what kind of amazing claims the salesman might have made. Tell me what you think it does."

"It's an electronic refrigerator," Herbie said. "You hook it up to electric power and it gets cold. I know you can't refrigerate something without power, but this uses electricity to cool stuff down."

Thermoelectric Coolers Are Hot

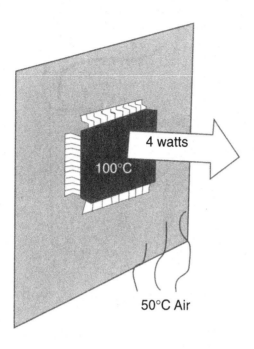

Figure 29-1 The INFERNO chip, with no heat sink in natural convection, is 50°C hotter than the incoming air

"That's close," I said, "but you don't think it sucks the heat out of components and converts it back to electricity, do you?"

Herbie chuckled, "No. That cheerleader guy who said that is not very technical. His job is to be optimistic, which is hard to do and still believe in the physical laws of the universe."

"OK," I said, "you know how a regular refrigerator works, like the one that makes ice for your bar in the basement. Left to itself, heat flows from hot to cold. But if you put in some power, such as by running a compressor, you can force heat to flow from cold to hot, and that way you can chill stuff down and make ice."

"OK," Herb nodded. "You put in power and make stuff cold."

"To be more precise, what you are doing is taking heat out of the stuff in your refrigerator and dumping it someplace else, such as into the air around your bar. It makes one place colder and another place

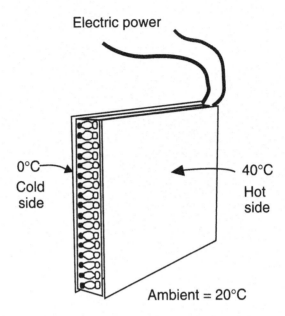

Figure 29-2 A thermoelectric cooler uses electricity to create a temperature difference

hotter," I said. "And a thermoelectric cooler does pretty much the same thing, without the milk jugs and novelty magnets holding up kids' fingerpaintings.

"A thermoelectric cooler is a sandwich. The bread is a couple of metal or ceramic plates, with a whole bunch of tiny thermocouples in parallel as the peanut butter in the middle. A thermocouple is just a pair of wires of different metals and if you heat one junction and cool the other, it produces a voltage. It also works in reverse. If you run a voltage from a power supply across the sandwich of thermocouples, one side gets colder than ambient, and the other side gets hotter than ambient." (Figure 29-2.)

Herbie asked, "So what's the problem? We stick the cold side against our component and it will make the component 0°C."

"It doesn't quite work that way," I said. "The thermoelectric cooler only makes a temperature *difference* between the hot and cold sides. The actual temperature of each side depends on the surround-

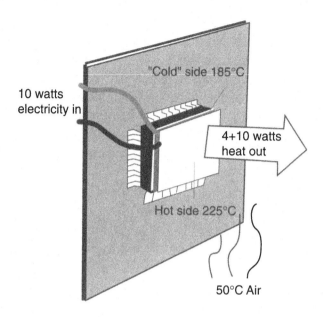

Figure 29-3 Unfortunately, the electricity turns into heat, which natural convection has to carry away, along with the original 4 watts from INFERNO

ings. Not only that, but the efficiency of the thermoelectric cooler is pretty darn low. To move 4 watts out of the INFERNO chip, it might take 10 watts of electric power to give you a 40°C temperature difference. That electric power turns to heat, too, and has to flow out of the hot side. So if you put a thermoelectric cooler on top of INFERNO, it would look like this." (Figure 29-3.)

"The airflow and geometry have not changed much from the situation you had with just the INFERNO chip. Remember that it got up to 100°C with only 4 watts dissipation. We are more than tripling the power, from 4 to 14 watts, so the surface temperature rise above the inlet air is going to increase more than three times. That makes the temperature on the hot side of the thermoelectric cooler about 225°C. The cold side is 40 degrees lower than that, a mild 185°C, and that is, of course, how hot the INFERNO chip would get," I finished.

Hot Air Rises and Heat Sinks

"I suppose if we didn't like the INFERNO chip at 100°C, then 185°C would be worse. So what good are these thermoelectric coolers?" Herbie whined in frustration.

"Even though they are costly and wasteful, there are some legitimate uses for them," I said. "I'll give you an example. Suppose you have to keep your chip at 40°C, but the ambient is at 50°C. No matter how big a fan or heat sink you throw at it, you can't get lower than the ambient. But if you put your thermoelectric cooler on the chip and *then* add a big heat sink, and blow on it with a fan so that the hot side is only 80°C, then you could actually cool the chip below ambient, down to 40°C. Laser transmitters for fiber optic communication use this idea. The laser diode doesn't work very well when it gets too hot, so they chill it down with a built-in thermoelectric cooler."

Herbie looked at his pile of viewgraphs and said, "I guess for the MMMnMM Project we'll take out the extra shelf and put the fan tray back in."

"Are you sure?" I asked. "I thought that once something was on a viewgraph, it was written in stone."

"True," Herb said, "but I'll file a VCR, a Viewgraph Change Request. I guess it was good you showed up to the meeting. Funny, but I don't remember inviting you. Oh, by the way, you've got some crumbs on the front of your sweater."

THE HOUSE OF CARDS

CHAPTER 30

> *Even the experts resort to a little mythologizing once in a while. A late night confession reveals that this business of controlling the temperature of electronics to improve performance and reliability isn't nearly as precise as is claimed. There is hope that someday soon the advance of technology might be able to slip a solid foundation under this house of cards without toppling the whole thing. Why isn't anybody worried?* **Lesson**: *The not-so-scientific relationship of temperature and reliability in electronics.*

"Heat transfer is not a precise art. If you can calculate something to within plus-or-minus 25%, you're doing a pretty good job."
— *Professor Bob Moffat, Stanford University, 1981.*

"All the latest research shows that if you stay below certain limits which are obviously destructive, the steady state temperature of an electronic component has very little to do with its long-term reliability."
— *Michael Pecht, University of Maryland, 1990.*

"You predicted my component will be at 127°C. The design guidelines say not to exceed 125°C. What happens if I don't fix the problem?"
— *Herbie, TeleLeap, 1997.*

It was one of those long nights in the lab. There was no compelling reason to stay late. The test results were not due anytime soon.

Hot Air Rises and Heat Sinks

A wintry twilight had settled over the parking lots around three in the afternoon and had stayed. Dozens of computer cooling fans smothered us in a nerve-dulling atmosphere of white noise as Herbie and I struggled to complete our latest temperature test.

The two-hour job had dragged on from lunch time until nearly 9 PM. A power lead would fall off, or the traffic-simulating software would lock up for no reason, forcing us to reboot the operating system. I spent an hour searching for a roll of printer paper for my thermocouple data logger, finally snagging one from the cafeteria cash register. It was the time of night when the vending machine coffee tastes particularly bitter. Life in general, and engineering projects in particular, seemed absolutely meaningless.

Herbie and I sat with our feet up on the lab bench, staring at the data logger LED display. It flashed the case temperature of a troublesome component on Herbie's prototype board, one he was not especially proud of. It was a redesign of one of TeleLeap's first products, a simple interface card for the original Crosser System (Crosser Classic™ in the sales brochure). It was so old that several of its components had been discontinued by the manufacturer. Herbie got the assignment of duplicating an old, slow, low-tech circuit with modern, available parts. The design work was not challenging, except for having to keep some of the original design mistakes that our customers had come to know as "features." No one was looking forward to the imminent release of this board. TeleLeap had to support the old Crossers in the field with replacement parts. We did not feel like rocket scientists that night.

The temperature test was supposed to have been a formality. The old interface board produced very little heat. However, because of the increase in component density since the original had been designed, 32 parts were now packed into one tiny package, and now this new transmission line driver chip was reading a tad on the warm side. Maybe it was curiosity that kept us in the lab, watching the line driver temperature creep up minute by minute.

At the end of another 10-minute interval, the thermal printer buzzed, spitting out thermocouple readings onto paper tape. Herbie ran the tape through his fingers like a 1920s stock tycoon. "The line

The House of Cards

driver hasn't changed in the last 20 minutes," he said. "Are you ready to give up and say that we've reached steady state?"

I would have agreed to almost anything. "Yeah, OK. Steady state. What did it turn out to be?"

Herbie punched some numbers on his hand calculator. "At the worst-case ambient of 50°C, I get that the junction temperature of this part will be — ummm — 127°C. That seems kind of high. Is that an OK junction temperature?"

I pulled out my pocket edition of the *TeleLeap Good Design Guidelines*. "It says here in our table of temperature limits that this kind of component should not be operated above 125°C, if we want to maintain good reliability."

Herbie let the paper tape drop. "Looks like I'm over the limit. I can't believe it! This is a no-brainer board, no high power, no high frequency, no nothing, and now you're going to make me redo the layout or add a heat sink or something. Are you really going to make me change my design at this point just for a crummy 2 degrees?"

I sighed, suddenly very tired. I asked, "Have you tested whether this board works right at the worst-case ambient?"

"Oh, yeah," Herbie said, "we had it in the chamber for several days at elevated ambient. It works fine, lots of timing margin and everything, all the way up to 70°C. It could probably go higher, but Doc needed the chamber for something important. And this board only has to work up to 50°C. We just need this test today to see if it meets the reliability requirements."

"In that case," I said, "I'll write the test report to say that the line driver exceeds that *G-D Guidelines* temperature limit, but the effect on reliability is slight, so I won't have to recommend any design changes."

Herbie was satisfied with that answer, and we started tearing down our setup. As I fished my thermocouples out of the prototype shelf, Herbie flicked off power switches all around the lab. I was thinking deeply and profoundly about which fast food drive-thru to hit for supper on the way home when Herbie stopped and sat down.

"But how can you just *say* that the risk is slight?" he asked. "Don't you at least have to give the temperature data to Sharon the

Reliability Engineer, so she can run it through her system reliability calculation program? For all of my other boards you made a big deal about how all my junction temperatures have to be under those limits in the *G-D Guidelines*. You made me add fans or heat sinks, or change the layout. I always thought that if we didn't, our customers would be sending back fried boards in little pine boxes in no time at all. And now you say for this board it's OK to be a couple of degrees over the limit. You act like these temperature limits don't mean anything at all!"

I unzipped the parka I had just thrown on and sat down across from him. There was something about the lateness of the hour, our solitude, the sleet pelting the windows, the long, frustrating job we had just shared that made it seem like a time for the truth. "OK," I said, letting out a long sigh. "If you have to know, then I'll tell you. The reason I can't get all worked up over a couple of degrees is that this whole business of controlling temperature in electronics is just a house of cards. We pretend we know a whole lot about how temperature affects the working of electronics, but mostly its a collection of procedures on top of misconceptions founded on myths.

"Let's start with our little test right here. You and I say that the junction temperature of that line driver will be 127°C when the ambient is 50°C. But how close do we really know that junction temperature? There are a few sources of error you ought to be aware of:

1. Thermocouple wire is calibrated to about plus-or-minus 2°C. That could pull your part under the limit all by itself. But the wire calibration is the least of my worries.

2. When you calculated the junction temperature just now from our case temperature measurement, you used the power dissipation of the line driver and the thermal resistance of the component package. I'll bet you a bag of bagels you don't really know the power dissipation of that line driver. You didn't measure it in the circuit — you used a worst-case number out of the data book. It might be off by 50% or more, either high or low. If we really wanted to know the junction temperature precisely, we'd have to measure the real power being turned into heat by the chip. But there's no point in doing all of that work because it would be masked by the error in the value of

R_{jc} — the thermal resistance between junction and case. No two vendors give the same number for the same kind of component package. The guys who wrote the industry standard for how to measure R_{jc} warn us not to put too much stock in the accuracy of it. And everyone admits that the value of R_{jc} depends on how fast the air is moving and how much copper is in the board. So to get our junction temperature, we take a thermocouple reading that could be off, and add to it a number generated by multiplying two guesses together.

3. Then there is the ambient. We did our test at the normal room temperature — about 23°C. To project what would happen if the ambient were 50°C, you just add 27 degrees to all of the measurements. This assumption is *probably* good most of the time. But does our test in this lab really duplicate the environment where the customer installs the Crosser? Maybe his or her air-conditioning vent won't be as close as this one over here. Maybe his 'still air' is not the same as our 'still air.' How repeatable is this test from one kind of room to another?

4. Components vary from lot to lot. Maybe the one we tested runs high in power dissipation, or maybe it is low. How many samples should we test to know we have got it properly bracketed? What about variations in all of the other parts on the board? Could they change the behavior of our line driver? How do they vary as the input voltage drifts, and what happens as the parts age and change characteristics over the 20-year life of the system?

5. Where did these operating temperature limits in the *G-D Guidelines* come from, anyway? I was told they were derived by somebody somewhere from actual field failure data. If that's true, how did that somebody know what the junction temperature was when the component failed in the field if we don't even have a good way of measuring junction temperature in the lab? Were those guesses any more precise than the ones we made tonight?

6. This line driver just came on the market last year. If our reliability limits are designed to help our circuits last for 20 years, how can anybody know what the real failure rate of a brand new component is? It hasn't existed for 10 or 20 years yet! How does anybody know

what temperature does to its long-term life? Ah, the magic words — *Accelerated Life Testing*. They run a batch of components at 200°C for six months and extrapolate the number of failures from that test down to what they think would happen at 70°C. Doesn't that sound just a little bit bogus? Well, it is! There are lots of failure mechanisms that only start happening way above 100°C. It's like life testing a candle above the melting point of wax and trying to use that test to say something about how it works at normal temperature. The technical articles I've seen (well, I looked at the abstracts at least) suggest that steady-state junction temperature doesn't do squat to long-term reliability.

"As far as I can tell, the operating temperature limits used in the electronics industry were pretty much pulled out of somebody's hat. We applied these temperature limits without a lot of justification, and the electronics seemed to work pretty well. Reliability and quality got better year after year, so nobody questioned the idea behind the temperature limits. Maybe controlling temperatures was a good thing, but was it just a coincidence that electronic assembly became more and more automated, so there were fewer and fewer human mistakes to make boards fail? If these limits are so scientific, doesn't it seem odd that the temperature limits used by the big companies like IBM and Motorola and Northrop and AT&T are all different from each other.

7. The last and most bogus: for decades we thermal gurus have been justifying our existence by claiming that if you control component temperature, you improve reliability. There was a very precise method for calculating just how much temperature affects the life of a component, a board, or a whole system. It was called MIL-HDBK-217, because it was put together by the military and their contractors. They were the folks who came up with that wonderful idea that every 10-degree rise in component temperature cut its life by half. This is what all of those reliability computer programs are based on, such as the one Sharon plugs and chugs with all day. The only problem with 217 is that it doesn't work. If you compare the failure rate predictions of 217 against real field failure rates, they look about as much alike as Madonna and Elvis. Even the Defense Department doesn't use it anymore."

The House of Cards

I twiddled the end of a thermocouple wire with the tip of a finger. I was sweating, and not just because of the winter coat. Herbie had his eyes half closed. He said, "I guess all of this goes double for the temperatures you get from a simulation."

I nodded. "Sure. Not even as good as a thermocouple test."

"So what you're saying is — let me see if I'm getting the whole story here — you can't really find out the junction temperature of a component, at least not to within a few degrees. And even if you could, nobody knows for sure what a good operating limit should be for the junction temperature. If somehow we managed to stay below that limit anyway, it probably didn't buy us any improvement in long-term reliability. Is that what you're saying?"

I squeezed the tip of the thermocouple between my thumb and forefinger. The 30-gage wire broke and the datalogger display lit up suddenly with "ALARM: OPEN T/C." "Yeah, Herb, that's about it," I said, hanging my head. I felt like Robert Preston at the end of *The Music Man*, when he's been unmasked as a flimflam artist. I partly regretted the verbal tar-and-feathering that I knew would come once Herbie had spread the word of my deception. But another part of me was relieved that the truth had come out at last.

"So what's the big deal?" Herbie said. "Why didn't you just tell me this stuff before? It's only *thermal*, after all. It's not like anybody cares about temperature. All we want you to do is tell us if the circuit board is going to work or not. We don't care if it's all that precise or scientific. Shake some chicken bones over the components, if that's what you have to do to make us think you know what you're talking about."

"What?"

Herbie snorted at me. "Look, I understand what's going on. You can't go around to everybody explaining that you've got a margin of error here, and you're not quite sure about this. Managers and customers want simple answers. You haven't lied to anybody — you just *simplified* a little.

"Everybody knows that heat is bad for electronics in some vague way. Last year, we had that power supply with thermal runaway, and then there was the clock circuit that started to drift when it got

Hot Air Rises and Heat Sinks

warm. If we didn't have a thermal watchdog like you, we would probably design our boards so hot that they'd glow like a toaster wire. So maybe you don't know *exactly* how hot my line driver junction temperature is, and you're not sure *precisely* what the operating limit should be. Somebody has to set some limits for us, or we'd go hog-wild and nothing would ever work. Let me tell you this: no matter how imprecise you think your thermal methods are, they are way more scientific than what we double-Es would come up with for ourselves, if we'd even bother until it was too late. So keep doing what you're doing. And if you have to throw in a little lie or exaggerate the precision of a measurement, or pull something out of your hat — do it! We need to believe that somebody has got a handle on this topic."

Herbie put his own coat on. My late night confession obviously did not have the same impact on him that it did on me. "So I haven't convinced you that my job is a house of cards that could come tumbling down?"

Herbie fished keys out of his pocket and said, "Remember that kid you brought in here a while back, the one from the Big Brothers program?"

"Matthew? Sure."

"You explained to him what you do all day long, showed him all this high-tech, blinky-light stuff. You told him you measure the temperature of electronics and figure out whether it's going to get too hot. Remember his reaction?"

"Yeah. He made a funny face and said, 'That's *it*?' "

Herbie was already headed down the hall, holding up one hand in farewell. "Keep that in mind. That is exactly how everybody else thinks about what you do." Then he was gone.

I nodded, knowing in my heart Herbie was right. I could probably build a whole career on the house of cards. It was good enough to get by. But I promised myself that I could do better than that, because I knew someday, maybe not too far in the future, I'd have to.

Herbie's Homework Helpers

> *If I have whetted your interest in heat transfer and cooling of electronics, or on the topic that everything you know in general is probably wrong, please go to these sources for lots more detail.*

Maybe you have come to realize that a lot of what you knew about cooling electronics was wrong. If this book has at least shaken your foundation a little bit, it has done its job. But I haven't replaced the myths, misconceptions, and mistakes with anything of substance. That I have left up to the pros, those who have been teaching and writing about heat transfer and cooling electronics for years. It would not be sporting of me to whet your appetite on this topic and then not at least give you a list of some of the finer restaurants that serve larger portions of this fare. You may find these works helpful if you get stuck solving your own thermal problems someday.

General References

Baumeister, Theodore, and Marks, Lionel S. (eds.). *Standard Handbook for Mechanical Engineers.* McGraw-Hill, New York. Get the latest edition. I am still using the one I got as a prize in college. It has very little to say directly about cooling electronics, but it is jam-packed with reference material, such as tables of thermal conductivity and properties of the atmosphere at high altitude. Make sure you have this, or another one like it, for looking up obscure things, especially when the libraries are closed.

Manual on the Use of Thermocouples in Temperature Measurement. ASTM STP 470, Philadelphia. This is way more information on thermocouples than I need, but it does have the calibration tables for various types of thermocouples. If you are going to get serious about making very accurate temperature measurements (down to 0.01°C), then this book will be useful.

Basic Heat Transfer

Holman, J. P. *Heat Transfer.* McGraw-Hill, New York, 1986. A popular (and readable!) college undergraduate heat transfer textbook. It is one of the few textbooks I bought *after* starting to work as an engineer. It is an excellent source for the basics of conduction, convection, and radiation heat transfer. It is not specific to cooling of electronics. The correlations about Nusselt numbers and isothermal ducts will not be of much direct use, but you have to understand the basics of heat and temperature before you can start to think about the thermal resistances between a silicon chip and an air stream. The tables of conversion factors and thermal properties of materials are extremely useful. I would buy this book again just for the appendices.

Kays, William M., and Crawford, Michael E. *Convective Heat and Mass Transfer.* McGraw-Hill, New York, 1980. This is a graduate-level textbook. You won't be able to read it on your own unless you already have a course or two in heat transfer and fluid dynamics under your belt. This is everything you always wanted to know about boundary layer theory and the Navier-Stokes equation, which is the basis of Computational Fluid Dynamics.

Moffat, Robert J. Course notes from "Experimental Methods in the Thermosciences." Bob Moffat is one of the few people on Earth who understands how thermocouples work. He shares this understanding, and other thermal experimental techniques, in a short course available periodically through the American Society of Mechanical Engineers (ASME).

Electronics Cooling

Kraus, Allan D., and Bar-Cohen, Avram. *Thermal Analysis and Control of Electronic Equipment.* McGraw-Hill, New York, 1983. This is a thorough introduction to the problem of cooling electronics. It is mainly qualitative and descriptive (it is not a textbook with problems to be worked). I found the section on optimizing finned heat sinks very useful in doing real design work.

Scott, Allan W. *Cooling of Electronic Equipment.* John Wiley & Sons, New York, 1974. This is a short, very practical book. It is not heavy on theory, but instead gives simple ways of calculating solutions to several examples of conduction cooling, air cooling, and radiation cooling. It is a good introductory book for someone who won't be doing thermal engineering full time. The edition I own is somewhat outdated (the example problems look like electronics from the Stone Age), but the heat transfer concepts are still valid.

Ellison, Gordon. *Thermal Computations for Electronic Equipment.* Krieger, Malabar, Florida, 1989. There is a lot of good stuff in here for the serious student of electronic cooling. The author has been solving all kinds of cooling problems in the electronic industry for decades. This book was his way of summarizing his problem-solving tools so that he could teach them to other engineers at his company. Especially useful, and not often found elsewhere, is his simple, practical, and even sufficiently accurate method for estimating the pressure drop for air flowing through electronic assemblies. The book comes with software that solves for temperature distribution and airflow using the resistance network method.

Pecht, Michael (ed.). *Handbook of Electronic Package Design.* Marcel Dekker, New York, 1991. To understand how heat moves through electric assemblies, you have to know how they are put together. This is a useful reference on electronic packaging. It has some material on cooling techniques. Mostly I use it to look up definitions of terms.

Lall, Pradeep; Pecht, Michael G; and Hakim, Edward B. *Influence of Temperature on Microelectronic and System Reliability.* CRC Press, Boca Raton, 1997. A summary of recent research efforts, this book is the cornerstone of the new understanding of the effects that temperature has on electronic reliability. It is the foundation of the new house of cards that is replacing the one that slid off of MIL-HDBK-217.

Industry Standards

Bellcore GR-63-CORE, "Network Equipment — Building System (NEBS) Requirements: Physical Protection." This is one of the many standards of the telecommunications industry. It describes the physical environment inside telephone central offices. Bellcore has standards for outdoor equipment and plenty of other things, too. If you work in the telecom field, this document will be a constant source of information and irritation for you. Unfortunately, it is not easy to get. Bellcore makes its living by selling copies of its documents, and they don't come cheap. If you just want to get a look at it, check around for a library that may have a copy on file.

UL 1950, "Safety of Information Technology Equipment, Including Electrical Business Equipment." Underwriters Laboratories. This is another industry standard that applies to telecom equipment (as well as to other business electronics). Here you will find the safety standard for how hot surfaces can be before they will burn the people trying to use them. Again, this document is only available from UL for a relatively large fee.

Technical Periodicals

To keep up with the latest trends in this field, take a look at these journals and magazines:

Journal of Electronic Packaging, published by the ASME (American Society of Mechanical Engineers).

Components, Packaging, and Manufacturing Technology, published by the IEEE (Institute of Electrical and Electronic Engineers).

Both of the above are peer-reviewed technical journals. Subscriptions are available at a reasonable rate (discounts to society members), or they are probably available in university libraries.

Electronics Cooling is a technical magazine aimed at the thermal engineer working in the electronics industry rather than in academia.

It has more articles about how to select a heat sink or use a CFD package than on a new solution approach to the Navier-Stokes equation. It is available by free subscription to qualified readers (people their advertisers are interested in reaching, I suppose). Contact them at 81 Bridge Road, Hampton Court, Surrey KY8 9HH, United Kingdom.

Getting the Right Attitude

Firesign Theatre, *Everything You Know Is Wrong*. This is a comedy album that first came out in 1975 and has been circulating in various bootleg forms ever since. Perhaps they will issue it on tape or CD someday. It is about "the South actually won the Civil War" and other things you don't know.

Gould, Stephen Jay. *The Mismeasure of Man*. W. W. Norton, New York, 1981. Gould has spent a career writing about science and making it accessible to the nonscientist, especially the field of biology and evolution. This book dismantles the popular notion (even among scientists) that people can be measured and valued by a single number, such as IQ. Not everything you read in a book or hear from a scientist is true.

Loewen, James W. *Lies My Teacher Told Me*. The New Press, 1995. This is a dead serious book about "everything your American history textbook got wrong." Read this and you'll be doubting the authenticity of your own birth certificate.

The Bible. Pick your favorite version. I pull out this book when none of the others work. Often the reason I can't see the solution to a problem is because my big head is in the way. Proverbs 3:7 advises, for example, "Never let yourself think that you are wiser than you are."

About the Author

Tony Kordyban picked up his knowledge of electronics cooling on the mean streets of such places as the University of Detroit, Stanford, and Bell Labs. Since 1990 he has been plying that trade full time at Tellabs, a leading developer and manufacturer of telecommunication equipment. This former cartoonist and wannabe novelist found an outlet for his creative juices there by starting up an informal newsletter on thermal design called *HOTNEWS*. It turned out that writing and illustrating children's stories for his family and friends was good practice for writing for engineers.

Although he has the credentials (BS and MS in Mechanical Engineering), he counts his 17 years of industrial experience (that is, of making lots of mistakes) as most valuable. Without mistakes he would have learned nothing, and had precious little to laugh about. And Tony thinks engineering is supposed to be fun.

INDEX

A

Accelerated life testing, 206
Air, "still," assumption of, 93
Air-cooled systems, 5
Air pressure, 75
Air speed, 144–145
Air temperature, 91–92, 188
 exit, 6
Air vents, 69
Airflow, 19, 106
 direction of, 124
Airflow disturbances, 125
Aluminum, 23, 39
"Ambient" temperature, 92–93, 205
Atmospheres of pressure, 76–77

B

Backup fans, 97–100
Baffles, 62
Barometers, 75–76
Bellcore standards, 212
Bible, 213
Bill of Material, 116

Blowers, 19
Boundary conditions, 33

C

Case, thermal resistance between junction and, 188, 193, 205
Case temperature, 94, 188, 192–193
Celsius conversion equation, 14
Centigrade conversion equation, 14
CFD (computational fluid dynamics), 33–37
CFM (cubic feet per minute), 54–57, 74
Cfmophobia, 54–57
Circuit, grounding, 112
Circuit boards, 3, 6
Clock frequency, 68
Clock speeds, 56
Component packages, heat paths inside, 189–193
Component power limits, 150–154
Component temperature, reducing, 48

217

Components
  heat from, 183–186
  high-power, 118
  junction temperature of, 188–193
  location of, 183–186
  maximum temperatures of, 50
  temperature of, 178–179
  temperature-sensitive, 118
  thermal resistance of, 89–90
  varying, 205
Computational fluid dynamics (CFD), 33–37
Computer thermal simulation, 115–120
Conduction, heat, 22–23
  through solids, 44–45
Conduction heat transfer, 29
Conductivity, 22–23, 24, 29–30
  units of, 26
Conservation of energy, 42
Constantan, 135
Convection, 16, 156
  forced, 16, 19, 146–147
  mixed, 146–147
  natural, see Natural convection
Convective heat transfer, 161–164
Convective heat transfer coefficient, 43, 102, 119
Conversion equation, Fahrenheit/Celsius, 14
Coolers, thermoelectric, 195–200
Cooling
  electronics, 8
  fan, 56
Cooling coils, 19
Cooling resources, 48
Cooling system, sieve, 64–66
Crosser system, 2, 3
Cubic feet per minute (CFM), 54–57, 74

D

Data books, 89–94
Data logger, 110, 111, 113
Delta triangle, 119
Diamond film substrates, 26
Dielectric test, 113
Diode temperature, 18–20

E

Echo Cancelers, 2
Electrical current sink, 41
Electrical energy, 42
Electrical isolation, 112
Electrical noise, 56
Electrical performance, thermal performance versus, 37
Electronics
  controlling temperature in, 204–208
  heat and, 207–208
Electronics cooling, 8
Electronics cooling books, 211–212
Energy, conservation of, 42
Energy balance equations, 34
Enviromatic 9000, 16–20
Environmental test chambers, 16–20
Epoxy
  silver, 22–24
  thermal, term, 25
Evaporation, 156
  humidity and, 106

Exit air temperature, 6
Extended surfaces, 40–41, 42
Extruded fin heat sink, 123–126

F

Fahrenheit conversion equation, 14
Fan cooling, 56
Fan curves, 77
Fans, 53–57, 144, 196
  backup, 97–100
  failure of, 96–100
  volumetric flow rate of, 74
Fin spacing, optimum, 125
Fluid dynamics, computational (CFD), 33–37
Forced convection, 16, 19, 146–147
Functional operating limits, 51

G

"G-D" Guidelines (*Good Design Guidelines*), 87–88
General reference books, 209–210
Good Design Guidelines ("G-D" Guidelines), 87–88
Gould, Stephen Jay, 213
Grounding circuit, 112

H

Heat
  from components, 183–186
  electronics and, 207–208
Heat conduction, *see* Conduction, heat
Heat conductors, 12
Heat energy, 29, 42

Heat flow, 106
  surface area and, 43
Heat flow rate, 30
Heat loss, rate of, 102
Heat path, 46
  inside component package, 189–193
Heat sink, 22, 39–46
  extruded fin, 123–126
  for homemade ZENO, 130–132
  ideal, 42
  pin fin, 46, 122–126
Heat spreaders, 25–26
Heat transfer
  conduction, 29
  convective, 161–164
Heat transfer equation, 104–106
Heat transfer technology, 8
Heat transfer textbooks, 210
Heating elements, 19
High-power components, 118
Hot spots, localized, 7
HOTNEWS, xiii
Human beings
  sweating of, 106, 156–157
  temperature limits of, 86–87
Humidity, evaporation and, 106

I

Industry standards, 212
Insulation, temperature ratings on, 83–84
Insulators, 131
  thermal, 24–26
Internet, 109
Isolation, electrical, 112

J

Joule monkeys, 27–30
Joules, 29, 30
Journals and magazines, 212–213
Junction
  term, 188
  thermal resistance between case and, 188, 193, 205
Junction temperature, 6
  of components, 188–193
  emphasis on, 188
  equation for, 188
  estimating, 94

K

Kapton tape, 111
Kordyban, Tony, 215

L

Landmarks on temperature scale, 10–14
Lasance, Clemens, 191–192
Laser transmitters, 200
Layouts, thermal, 33–37
Leakage paths, 131–132
LED junctions, 50–51
Life testing, accelerated, 206
Localized hot spots, 7

M

Magazines and journals, 212–213
Margin test, traditional, 20
Mass balance equations, 34
Mass storage devices, 179–180
Maximum operating limits, 49–51
Maximum operating temperature, 11

Metric system, 9
Metrics, 47
Microprocessors, power
  dissipation of, 68–72
  reducing, 72
MIL-HDBK-217, 49, 206
Misdata books, 89–94
Mixed convection, 146–147
Moffat, Bob, 201, 210
Multiplexing, 3

N

Natural convection, 16–17, 19, 20, 145–147
  in air, 69, 70
  calculating, 34
  defined, 35
  power density in, 32
  practical limit of, 68–72
  in stacks, 61–66
Natural convection power limits, 153–154
Noise, electrical, 56

O

Ohm's Law, 189, 190
Operating limits
  functional, 51
  maximum, 49–51
Optical disk drives, 175–180

P

Paper tape punch, 179–180
PCBs (printed circuit boards), 11
Pecht, Michael, 201
Peltier effect, 110

Periodicals, technical, 212–213
Pessimism, 96
"Phone tag," 31
Pin fin heat sink, 46, 122–126
Pinky Rule, 126
Power, 30, 120
  typical, 119
Power density in natural convection, 32
Power dissipation, 153, 204
  of microprocessors, 68–72
  reducing, 72
Power limits
  natural convection, 153–154
  thermal, 150–154
Pressure
  air, 75
  atmospheres of, 76–77
Printed circuit boards (PCBs), 11

R

Reference books, general, 209–210
Reference points on temperature scale, 10–14
Refrigerators, 196–198
Reliability, 6, 203
  improving, 97
  maximizing, 49
Reliability limits, 205–206
Reliability temperature limits, 49–51
Resistance
  electrical, 26
  thermal, see Thermal resistance
Resistance network, 191–192

Root cause analysis, 162, 164
Rosten, Harvey, 191–192
Rule of Thumb, 31–32, 122
Runaway, thermal, 19–20

S

Safety switch, thermal, 99
Seebeck effect, 110
Set-point temperature, 19
Shut down, thermal, 99
Sieve cooling system, 64–66
Silicon feature sizes, 68
Silver epoxy, 22–24
Simulation, computer thermal, 115–120
Sink, 41
  heat, see Heat sink
Skin temperature, 102, 158–159
Solar radiation, 92
Solder, 142
  melting, 11
Speed limit, 151
Spider-web resistance method, 191–192
SportsFan system box, 144–145
Steady state temperature, 167–173
"Still air," assumption of, 93
Styrofoam, 23
Sunscreen, 155–159
Surface area, 124
  exposed, 106
  extended, 42
  heat flow and, 43
Surface temperature, 83, 199
Sweating, human, 106, 156–157

T

Tape cartridges, 179
Technical periodicals, 212–213
TeleLeap, 1–3
Telepathy, 73–74
Temperature, 29
  air, *see* Air temperature
  "ambient," 92–93, 205
  case, 94, 188, 192–193
  of components, 178–179
  reducing, 48
  controlling, in electronics, 204–208
  diode, 18–20
  exit air, 6
  junction, *see* Junction temperature
  maximum, of components, 50
  maximum operating, 11
  set-point, 19
  skin, 102, 158–159
  steady state, 167–173
  surface, 83, 199
  transformer, 172–173
  voltage and, 110, 111–112
  windchill, 104–107
Temperature difference, 22–23, 119
Temperature limits
  of human beings, 86–87
  reliability, 49–51
Temperature-measuring watch, 9, 90–91
Temperature ratings on insulation, 83–84
Temperature rise, 46

Temperature scale(s)
  landmarks on, 10–14
  relative, 157–159
Temperature-sensitive components, 118
Temperature test reports, 4–5
Temperature tests, 202–208
Test chambers, environmental, 16–20
Thermal conductivity, *see* Conductivity
Thermal design goals, 80–88
Thermal epoxy, term, 25
Thermal insulators, 24–26
Thermal layouts, 33–37
Thermal performance, electrical performance versus, 37
Thermal power limits, 150–154
Thermal resistance
  of components, 89–90
  between junction and case, 188, 193, 205
Thermal runaway, 19–20
Thermal safety switch, 99
Thermal shut down, 99
Thermal simulation, computer, 115–120
Thermal time constant, 168–171
ThermaNator, 59, 127
  as test equipment, 127–128
Thermocouple junctions, 111, 133–142
Thermocouples, 109–113, 198
  insulating, 111
Thermoelectric coolers, 195–200
Thermotriples, 135–138

Time constant, thermal, 168–171
Traditional margin test, 20
Transformer temperature, 172–173
Transformers, 171
Transistors, 189

U

Underwriters Laboratories
    standards, 212

V

Van Allen belts, 106
Vents, air, 69
Voltage
    reducing, 72
    temperature and, 110, 111–112
Volumetric flow rate of fans, 74

W

Wall, The, 68–72
Watts, 30, 181–186
    defined, 181–182
Windburn factor, 105–106
Windchill, idea of, 102
Windchill factor, 101–107
Windchill temperature, 104–107
Wiretap devices, 60

Z

ZENO custom chip, 11
    homemade, 128–132

CPSIA information can be obtained
at www.ICGtesting.com
Printed in the USA
BVOW04s2018111116
467351BV00009B/118/P